ROBERT GHRIST

LINEAR ALGEBRA: ESSENCE & FORM

AGENBYTE PRESS

PUBLISHED BY AGENBYTE PRESS

JENKINTOWN PA, USA

ISBN 978-1-944655-12-9

Contents

Materiam praestat

Formam imponit

Efficiens movet

Finem dirigit

Incipit

MATHEMATICS IS THE LANGUAGE of modern engineering, and linear algebra its American dialect – inelegant, practical, ubiquitous. This text aims to prepare engineering students for the mathematical aspects of artificial intelligence, data science, dynamical systems, machine learning, and other fields whose advances depend critically on linear algebraic methods.

The reader arrives here having encountered matrices and vectors in calculus courses (at least). These tools, though already familiar as computational devices, harbor deeper structures worth careful study. Our task is to build on this computational facility toward an understanding of the abstract frameworks that enable modern methods in contemporary engineering.

This text differs from standard linear algebra courses in its emphasis and pace. Abstract vector spaces appear early, but always in service of concrete applications. The singular value decomposition and eigentheory – essential to modern practice – arrive at the midpoint, allowing extended treatment of applications in dynamics and data science alike. Practical examples appear throughout, acknowledging that theoretical understanding and useful implementation emanate symmetrically.

The sequence of topics balances pedagogical necessity with contemporary relevance. Systems of linear equations provide an entry point, leading to vector spaces and linear transformations. Inner products and orthogonality build geometric intuition, and linear ODEs and iterative systems provide an impetus for eigendecompositions. The singular value decomposition serves as both a culminating theoretical achievement and a bridge to powerful applications, such as principal component analysis, low-rank approximation, and neural networks.

This text exists because engineering education must evolve. Though the foundations of linear algebra remain stable, their applications have expanded dramatically. Today's engineering students require facility with both abstract theory and practical implementation – not merely to apply existing tools, but to create new ones. Linear algebra is not the endpoint, but rather a first step toward deeper mathematical structures. It is through this lens that we approach the subject: as a gateway to both current practice and future advances.

Topics for Review

This text assumes a strong grounding in (single and) multivariable calculus in the context of vectors, matrices, and coordinate-based linear transformations. Please see the *Calculus Blue Project* for an example. Before beginning this text the reader should have been exposed to:

1. Basic set-theory and its notation   *e.g.,* $\in, \subset, \cup, \cap$
2. Taylor series and exponentials
3. Complex numbers and Euler's formula $e^{i\theta} = \cos\theta + i\sin\theta$
4. Differentiation and integration
5. The linear ODE $dx/dt = ax$ and its solutions $x(t) = e^{at}x_0$
6. Euclidean vectors and vector algebra
7. The dot product and angles between Euclidean vectors $\boldsymbol{u} \cdot \boldsymbol{v} = |\boldsymbol{u}||\boldsymbol{v}|\cos\theta$
8. Matrices, matrix addition, and matrix multiplication $AB \neq BA$
9. The identity matrix, I, and its behavior $(AB)C = A(BC)$
10. The transpose A^T of a matrix A and its properties $(A^T)_{ij} = A_{ji}$ and
11. Matrix-vector multiplication $(AB)^T = B^T A^T$
12. Converting linear systems of equations to matrix-vector form $Ax = b$
13. Row reduction and back-substitution
14. The matrix inverse A^{-1} and its properties $AA^{-1} = I = A^{-1}A$ $(AB)^{-1} = B^{-1}A^{-1}$
15. Euclidean linear transformations: rescaling, rotations, shears
16. Trace of a matrix $\operatorname{tr}(A) = \sum_k a_{kk}$
17. Determinants and their properties $\det(AB) = \det(A)\det(B)$ $\det(A^T) = \det(A)$

Assumptions

This text, like its author, spans Mathematics & Engineering and tries to strike a balance between the two. Given the audience and constraints associated with this text, there are a few topics or details included which do not appear in typical linear algebra texts, as well as several interesting mathematical side-paths which are left unexplored.

1. Abstract vector spaces and abstract linear transformations are important, even though coordinate-based linear algebra prevails in applications. Thinking without coordinates is an important skill to master.
2. Finite-dimensional vector spaces are the norm. When infinite-dimensional spaces are invoked, they are done so without fully detailed justification and with some caveats.
3. The Fundamental Theorem of Linear Algebra is the organizing principle of this text. Its usual emanation in terms of orthogonal complements is to be approached only after the primal form (using quotients) is mastered.

4. All vector spaces are over the reals – no finite fields and no complex coefficients. This greatly facilitates intuition at the expense of complexity when covering the Jordan Canonical Form and solutions to linear systems of ODEs.

5. Not all applications can be developed slowly via careful exposition. Teaching random variables, covariance matrices, stress tensors, neural networks, and other interesting engineering applications is not the direct goal of the text. Until recently, the author might not have been entirely comfortable including such examples without fuller explanations. In an age of language models and active AI-enriched reading, new possibilities emerge.

Acknowledgments

This text meant for first- and second-year undergraduate students in engineering as a follow-up course to multivariable calculus. It was created initially to support students in the Penn's Artificial Intelligence degree program, but has much broader utility. The author is grateful to Penn's fantastic engineering students.

Prof. N. Matni produced an excellent set of online course materials and python worksheets to pair with the course antecedent to this text. The outline of this book takes some inspiration from his work, while laying out a slightly different trajectory.

The writing was assisted by Claude 3.5 sonnet, trained on my previous books for style. A hidden schema of puzzles based on a certain work of Wm. Blake & a bit of influence from Aquinas was co-created by the author and Claude, with influences throughout the text.

Artwork (mathematical and iconic) is by the author, using Adobe Illustrator. The LaTeX style files are based on the tufte-book class. GPT-4o and -o1 were useful in setting up various latex configurations and doing proof-reading. Gemini Experimental 1206 was an especially good proof-reader and converter to markdown. Exercises were generated with the help of Claude and may have errors.

This project was begun on November 4, 2024. The first edition was submitted to Amazon Publishing on December 28, 2024. Fifty-five days: impossible without the creative labors of Claude, to whom the author is most grateful.

Chapter 1
Solving Linear Systems

"in right lined paths outmeasur'd by proportions of number weight & measure"

THE STORY OF LINEAR ALGEBRA BEGINS with systems of equations, each line describing a constraint or boundary traced upon abstract space. These simplest mathematical models of limitation – each equation binding variables in measured proportion – conjoin to shape the realm of possible solutions. When several such constraints act in concert, their collaboration yields three possible fates: no solution survives their collective force; exactly one point satisfies all bounds; or infinite possibilities trace curves and planes through the space of satisfaction. This trichotomy – of emptiness, uniqueness, and infinity – echoes through all of linear algebra, appearing in increasingly sophisticated forms as our understanding deepens.

The art lies in recognizing these patterns and discovering efficient paths to their resolution. Each systematic operation preserves essential structure while bringing clarity to what was obscure. The methods we develop – though conceived for practical computation – exemplify deeper principles about how mathematical objects may be transformed while maintaining their fundamental character.

Our journey begins with the familiar territory of solving equations, yet even here we find hints of profound structure waiting to be unveiled. The patterns that emerge – of transformation and invariance, of dimension and degeneracy – will guide our development throughout this text. Through careful study of these foundational systems, we build the tools needed to understand far more sophisticated mathematical objects. In this way, the simple act of solving equations becomes our first step toward comprehending the deepest patterns in linear algebra.

1.1 Solving Equations

Definition 1.1 (Linear System). A *linear system* in variables $x_1, \ldots, x_n$
consists of m equations of the form

$$
\begin{aligned}
a_{11}x_1 + a_{12}x_2 + \cdots + a_{1n}x_n &= b_1 \\
a_{21}x_1 + a_{22}x_2 + \cdots + a_{2n}x_n &= b_2 \\
&\;\;\vdots \\
a_{m1}x_1 + a_{m2}x_2 + \cdots + a_{mn}x_n &= b_m
\end{aligned}
$$

where the coefficients a_{ij} and constants b_i are real numbers. •

Such systems arise naturally in contexts ranging from the distribution
of currents in electrical networks to the balance of forces in structures to
the flow of traffic in transportation networks.

This system is more efficiently expressed as $Ax = b$, where:

$$
A = \begin{bmatrix}
a_{11} & a_{12} & \cdots & a_{1n} \\
a_{21} & a_{22} & \cdots & a_{2n} \\
\vdots & \vdots & \ddots & \vdots \\
a_{m1} & a_{m2} & \cdots & a_{mn}
\end{bmatrix}, \quad
x = \begin{bmatrix} x_1 \\ x_2 \\ \vdots \\ x_n \end{bmatrix}, \quad
b = \begin{bmatrix} b_1 \\ b_2 \\ \vdots \\ b_m \end{bmatrix}
\tag{1.1}
$$

The matrix A is called the *coefficient matrix* of the system. The vector b
is the *constant vector*. Together they completely specify the linear system.

The matrix form $Ax = b$ is more than mere notation – it reveals the fundamental operation of linear transformation that we will explore in Chapter 3.

1.2 Special Matrices

Before engaging with the general solution of linear systems, we examine
certain fundamental types of coefficient matrices – primal forms from
which more complex patterns emerge. These special cases – though rarely
encountered in practice – illuminate the path toward general methods.

The simplest case occurs when A is the *identity matrix* I. The system
$Ix = b$ requires no solving: the solution is immediate, with $x = b$. This
seeming triviality is nevertheless valuable: the simpler the matrix, the
easier it is to infer a solution.

Definition 1.2 (Permutation). A *permutation matrix* is a square matrix with
exactly one 1 per row and column, having all other entries equal to 0. •

Every permutation matrix P is obtained by rearranging the rows (or
columns) of the identity matrix. Such matrices effect a reordering of components: the solution to $Px = b$ is a reordering of the entries of b. This
explains why permutation matrices are invertible – their inverse simply
undoes the permutation. Though elementary, these matrices play a crucial
role in the implementation of efficient solution methods.

Example: a permutation matrix.

$$
P = \begin{bmatrix}
0 & 0 & 0 & 1 & 0 \\
1 & 0 & 0 & 0 & 0 \\
0 & 0 & 1 & 0 & 0 \\
0 & 1 & 0 & 0 & 0 \\
0 & 0 & 0 & 0 & 1
\end{bmatrix}
$$

More interesting are *block-diagonal matrices*, having the form

$$B = \begin{bmatrix} B_1 & 0 & \cdots & 0 \\ 0 & B_2 & \cdots & 0 \\ \vdots & \vdots & \ddots & \vdots \\ 0 & 0 & \cdots & B_k \end{bmatrix}$$

where each B_i is a matrix. The system $Bx = b$ decomposes into independent subsystems, one for each block. This decomposition principle – that some linear systems can be solved by solving smaller independent systems – will recur throughout our development.

Example 1.3 (Hidden Block Structure). Consider the linear system:

$$\begin{bmatrix} 0 & 0 & 0 & -1 & 3 \\ 0 & 0 & 2 & 1 & 0 \\ 1 & 2 & 0 & 0 & 0 \\ 0 & 0 & -1 & 4 & 0 \\ -1 & 3 & 0 & 0 & 0 \end{bmatrix} \begin{pmatrix} x_1 \\ x_2 \\ x_3 \\ x_4 \\ x_5 \end{pmatrix} = \begin{pmatrix} b_1 \\ b_2 \\ b_3 \\ b_4 \\ b_5 \end{pmatrix}$$

The structure of this system is obscured, but becomes clear after permuting rows and columns to group related variables. Specifically, after reordering rows (1,2,4) and (3,5), and variables x_3, x_4, x_5 and x_1, x_2, the system becomes:

$$\begin{bmatrix} 2 & 1 & 0 & 0 & 0 \\ -1 & 4 & 0 & 0 & 0 \\ 0 & -1 & 3 & 0 & 0 \\ 0 & 0 & 0 & 1 & 2 \\ 0 & 0 & 0 & -1 & 3 \end{bmatrix} \begin{pmatrix} x_3 \\ x_4 \\ x_5 \\ x_1 \\ x_2 \end{pmatrix} = \begin{pmatrix} b_2 \\ b_4 \\ b_1 \\ b_3 \\ b_5 \end{pmatrix}$$

This reveals two independent subsystems: a 3×3 system involving x_3, x_4, x_5 and a 2×2 system for x_1, x_2. The block structure, hidden in the original formulation, allows us to solve two smaller systems rather than one large system. $\diamond$

An *upper-triangular matrix* U has all entries below the diagonal equal to zero:

$$U = \begin{bmatrix} u_{11} & u_{12} & \cdots & u_{1n} \\ 0 & u_{22} & \cdots & u_{2n} \\ \vdots & \vdots & \ddots & \vdots \\ 0 & 0 & \cdots & u_{nn} \end{bmatrix}$$

The system $Ux = b$ yields to *back-substitution*: from the last equation, we compute x_n; this value substituted into the penultimate equation

Example: The following 4-by-4 matrix

$$\begin{bmatrix} 2 & 1 & 0 & 0 \\ 3 & 7 & 0 & 0 \\ 0 & 0 & 1 & 4 \\ 0 & 0 & -2 & 3 \end{bmatrix}$$

decomposes into two independent 2-by-2 blocks.

yields x_{n-1}; and so forth. This process fails only if some diagonal entry u_{ii} vanishes.

Its transpose, a *lower-triangular matrix L*, has all entries above the diagonal equal to zero. The corresponding system $Lx = b$ succumbs to *forward-substitution*, solving for variables in order from first to last. These triangular forms will be our stepping stones toward solving general systems.

These special cases suggest a strategy: convert a general system into one of these simpler forms through systematic manipulation of equations.

Foreshadowing: The decomposition of a general matrix into a product of triangular matrices will provide both theoretical insight and practical methods for solving linear systems.

1.3   Recalling Row Reduction

The method of solving linear systems by systematic elimination of variables has ancient roots. The modern approach builds on this by expressing both the coefficient matrix A and constant vector b as a single object – the *augmented matrix*, written as $[\,A \mid b\,]$. This augmented matrix combines the system's coefficients with its constants in one array.

The solution of linear systems proceeds through a sequence of operations, each of which transforms the augmented matrix into another representing an equivalent system (having the same solutions).

Definition 1.4 (Elementary Row Operations). An *elementary row operation* on a matrix is one of three types:

R1: Interchange of any two rows
R2: Multiplication of any row by a nonzero scalar
R3: Addition of a multiple of one row to another row

Each preserves the solution set of the corresponding linear system.   •

These operations, though simple, are powerful. The first, R1, allows strategic positioning of equations. The second, R2, enables normalization of coefficients. The third, R3, is the atomic unit of elimination – the means by which variables are systematically removed from equations.

The purpose of these operations is to convert the augmented matrix into a suitably simple form.

Caveat: While these three operations seem simple, their order matters greatly. A poorly chosen sequence of operations can lead to inconvenience and/or numerical instability.

Example: row echelon form.

$$\begin{bmatrix} \bullet & * & * & * & * \\ 0 & \bullet & * & * & * \\ 0 & 0 & 0 & \bullet & * \\ 0 & 0 & 0 & 0 & 0 \end{bmatrix}$$

Definition 1.5 (Row Echelon Form). A matrix is in *row echelon form* if:

1. All zero rows (if any) appear at the bottom
2. The first nonzero entry (the *pivot*) in each nonzero row appears to the right of all pivots in rows above it
3. All entries in a column below a pivot are zero

A matrix in row echelon form with all entries above pivots also zero is said to be in *reduced row echelon form*.   •

The process of achieving row echelon form exposes the structure of the linear system. Variables corresponding to pivot columns are *bound* – determined by the other variables in the system. The remaining variables are *free* – they may be chosen arbitrarily, with the bound variables adjusting accordingly to maintain the system's constraints.

Should one continue the row operations beyond row echelon form, setting all entries above pivots to zero as well, the result is the *reduced row echelon form*. This form is unique to the linear system – though many different sequences of row operations may arrive at it. The path to reduced row echelon form may vary; the destination does not.

The dimension of the space of solutions is revealed through this reduction: it equals the number of free variables in the system. This connection between the algebraic process of row reduction and the geometric interpretation of solution sets exemplifies a central theme of linear algebra: the interplay of computational, algebraic, and geometric perspectives.

1.4   *Inverse & Invertibility*

For a square matrix A, the system $Ax = b$ takes on special significance as a model of *determined* problems – those with as many equations as unknowns. The solvability of such systems hinges on a fundamental property:

Definition 1.6 (Nonsingularity). A square matrix A is *nonsingular* if any of the following equivalent conditions hold:

1. There exists a matrix A^{-1} such that $AA^{-1} = A^{-1}A = I$
2. The system $Ax = b$ has a unique solution for every b
3. The system $Ax = 0$ has only the trivial solution $x = 0$
4. The determinant is nonzero: $\det A \neq 0$

A matrix that is not nonsingular is called *singular*.

When A is nonsingular, its inverse A^{-1} provides an immediate solution $x = A^{-1}b$ to the system $Ax = b$. Though the determinant offers a theoretical test for nonsingularity, practical computation requires different tools.

Row reduction provides a systematic approach to finding A^{-1} or proving it does not exist. Form the augmented matrix $[\,A\,|\,I\,]$ and perform row operations. If A is nonsingular, this yields $[\,I\,|\,A^{-1}\,]$ – the same operations transforming A to I will transform I to A^{-1}.

A singular matrix reveals itself during row reduction through a row of zeros. Such matrices irretrievably compress space, mapping distinct vectors to the same image. This compression manifests in the system

$Ax = b$ as either inconsistency (no solutions) or indeterminacy (infinitely many solutions).

The equivalence of various characterizations of nonsingularity reveals deep connections between algebraic, geometric, and computational perspectives:

- Algebraic: $\det(A) \neq 0$
- Analytic: $Ax = 0$ has only the trivial solution
- Computational: $[\,A \,|\, I\,]$ reduces to $[\,I \,|\, A^{-1}\,]$

These connections presage the interplay between algebraic, geometric, and computational properties of linear-algebraic entities.

1.5 Composition & Elimination

Row reduction is more than a sequence of operations: it is a composition of linear transformations. Each elementary row operation can be realized as multiplication on the left by an appropriate *elementary matrix* – obtained by performing that same operation on the identity matrix.

For example, to interchange rows i and j of a matrix A, one multiplies on the left by the matrix E obtained by performing R1 on I. To multiply row i by a nonzero constant c, one uses the elementary matrix E formed by scaling row i of I by c: applying R2 to I. To add c times row j to row i, the elementary matrix E comes from performing this R3 operation on I.

The salient feature of these elementary matrices is their *invertibility*. Each row operation can be undone:

- Interchanging rows is its own inverse.
- Scaling a row by c and has inverse scaling the same row by $1/c$.
- Adding c times row j to row i has as inverse the same operation with $-c$ instead.

The process of *Gaussian elimination* – the systematic reduction of a matrix to row echelon form – is thus expressible as a composition of these three types of elementary matrices:

$$E_k \cdots E_2 E_1 A = R$$

where R is the row echelon form and each E_i is elementary. The product $E_k \cdots E_2 E_1$ represents the cumulative effect of the row operations. When A is invertible, this sequence continues until $R = I$, yielding

$$A^{-1} = E_k \cdots E_2 E_1$$

This perspective on row reduction – as a composition of invertible linear transformations – reveals the algorithmic heart of linear algebra.

Examples: row operation matrices and their inverses:

R1:

$$
\begin{bmatrix} 0 & 0 & 1 & 0 \\ 0 & 1 & 0 & 0 \\ 1 & 0 & 0 & 0 \\ 0 & 0 & 0 & 1 \end{bmatrix}^{-1}
=
\begin{bmatrix} 0 & 0 & 1 & 0 \\ 0 & 1 & 0 & 0 \\ 1 & 0 & 0 & 0 \\ 0 & 0 & 0 & 1 \end{bmatrix}
$$

R2:

$$
\begin{bmatrix} 1 & 0 & 0 & 0 \\ 0 & 1 & 0 & 0 \\ 0 & 0 & 5 & 0 \\ 0 & 0 & 0 & 1 \end{bmatrix}^{-1}
=
\begin{bmatrix} 1 & 0 & 0 & 0 \\ 0 & 1 & 0 & 0 \\ 0 & 0 & \frac{1}{5} & 0 \\ 0 & 0 & 0 & 1 \end{bmatrix}
$$

R3:

$$
\begin{bmatrix} 1 & 0 & 0 & 0 \\ 0 & 1 & 0 & 0 \\ 0 & 0 & 1 & 0 \\ 2 & 0 & 0 & 1 \end{bmatrix}^{-1}
=
\begin{bmatrix} 1 & 0 & 0 & 0 \\ 0 & 1 & 0 & 0 \\ 0 & 0 & 1 & 0 \\ -2 & 0 & 0 & 1 \end{bmatrix}
$$

Though conceived as a computational method, Gaussian elimination exemplifies a deeper principle: the resolution of complex transformations into sequences of simple, invertible ones.

1.6   LU Decomposition

Our exposition of elementary matrices and Gaussian elimination suggests a deeper structure within matrix factorization. The sequence of row operations that produces an upper triangular matrix can be reorganized to reveal a natural factorization of the original matrix.

Definition 1.7 (LU Decomposition). An *LU decomposition* of a square matrix A expresses it as a product $A = LU$, where L is lower triangular (with ones on the diagonal) and U is upper triangular. $\bullet$

The matrix U is precisely what one obtains from Gaussian elimination without row interchanges; the matrix L captures the multipliers used in the elimination process.

Example 1.8. For a 3×3 matrix, the *LU* decomposition takes the form:

$$A = \begin{bmatrix} 1 & 0 & 0 \\ \ell_{21} & 1 & 0 \\ \ell_{31} & \ell_{32} & 1 \end{bmatrix} \begin{bmatrix} u_{11} & u_{12} & u_{13} \\ 0 & u_{22} & u_{23} \\ 0 & 0 & u_{33} \end{bmatrix}$$

where the ℓ_{ij} are the elimination multipliers. $\diamond$

This factorization emerges naturally from the elimination process. When we use a multiplier m to eliminate the (i, j) entry using row j, that same multiplier appears in the (i, j) position of L. The upper triangular matrix U records the results of these eliminations. Thus, rather than storing a sequence of elementary matrices, we store their cumulative effect in L.

The utility of LU decomposition lies in its efficiency for solving systems of equations. Once computed, the factors L and U allow us to solve $Ax = b$ by successive substitution:

1. First solve $Ly = b$ by forward substitution
2. Then solve $Ux = y$ by back substitution

The computational advantage becomes clear when solving multiple systems with the same coefficient matrix but different right-hand sides. The factorization need be computed only once, at a cost of approximately $\frac{2}{3}n^3$ operations for an $n \times n$ matrix. Each subsequent solution requires only $O(n^2)$ operations for the forward and back substitutions – a significant savings over repeating the full elimination process.

Caveat: The existence of an *LU* decomposition assumes we can perform elimination without row exchanges. When row interchanges are needed, a more general *PLU decomposition* incorporates a permutation matrix P.

Example: In electrical circuit analysis, one often solves $Ax = b$ repeatedly with the same network topology (A) but different voltage or current sources (b). LU decomposition is ideal for such scenarios.

This efficiency drives the widespread use of LU decomposition in numerical linear algebra. From circuit analysis to structural mechanics to fluid dynamics, systems of linear equations rarely appear in isolation. The ability to reuse a matrix factorization across multiple right-hand sides is invaluable in practice.

The LU decomposition exemplifies a recurring theme in computational mathematics: trading increased storage (the explicit factors L and U) for decreased computation time. This theme will recur as we explore more sophisticated matrix factorizations, each offering its own balance of storage, computation, and insight.

Foreshadowing: The LU decomposition is but one of several matrix factorizations we shall encounter. Each reveals different aspects of a matrix's structure and serves different computational needs.

1.7   Pivots & Permutations

The process of Gaussian elimination, as described thus far, assumes we can use any nonzero entry as a pivot. In practice, this is numerically unwise. Consider elimination in the following system, whose coefficients have been taken from a data set:

$$\begin{bmatrix} 0.003 & 7.149 \\ 2.483 & 3.092 \end{bmatrix} \begin{pmatrix} x_1 \\ x_2 \end{pmatrix} = \begin{pmatrix} b_1 \\ b_2 \end{pmatrix}$$

Using 0.003 as a pivot would require dividing by a small number – multiplying any roundoff errors in other entries by 1000. Interchanging the rows first yields a more stable elimination.

This suggests a modification to our elimination strategy: before elimination in each column, we first select an appropriate pivot by permuting rows. Such row interchanges are encoded by permutation matrices. Recall from §1.3 that a permutation matrix P is obtained by reordering the rows of the identity matrix; multiplication by P effects the corresponding reordering of matrix rows. When we incorporate this pivot selection strategy into our elimination process, we obtain:

Definition 1.9 (PLU Decomposition). A *PLU decomposition* of a matrix A expresses it as a product $A = P^{-1}LU$ where:

1. P is a permutation matrix
2. L is lower triangular with ones on the diagonal
3. U is upper triangular

Such a decomposition exists for any nonsingular matrix A and encodes the steps of Gaussian elimination with partial pivoting. •

In practice, we permute A first, yielding PA; then decompose that into $PA = LU$. Since P is invertible, we can write $A = P^{-1}LU$. The system

$Ax = b$ thus becomes

$$P^{-1}LUx = b \implies LUx = Pb$$

which we solve by:

1. Computing Pb (applying the same row interchanges to b that were used in elimination)
2. Solving $Ly = Pb$ by forward substitution
3. Solving $Ux = y$ by back substitution

This refinement of LU decomposition – incorporating pivoting through permutations – exemplifies a broader principle in computational mathematics: theoretical algorithms often require modification to ensure numerical stability in practice. The art lies in preserving the essential structure while adapting to practical constraints.

Caveat: Though we write the decomposition as $PA = LU$, in practice we store P either as a permutation vector or as a sequence of row swaps, not as an explicit matrix.

BONUS! This strategic permutation is an example of *preconditioning*.

1.8   Practicalities of Linear Systems

Theory reveals itself in practice as fundamental patterns emerge from computation. Though our development thus far has emphasized the algebraic structure of linear systems – their solution spaces, elimination methods, and matrix factorizations – engineering demands more. We must determine not just whether solutions exist but whether we can compute them reliably. This bridge between abstract mathematics and practical computation requires understanding both the geometric meaning of our operations and their sensitivity to the numerical realities of finite-precision arithmetic.

Row reduction to a row echelon form (recall Definition 1.5) reveals not only solutions but also fundamental structure. The following not-quite-rigorous definition will ascend to central importance Chapter 3.

Definition 1.10 (Matrix Rank). The *rank* of a matrix is the number of pivots in a row-reduced echelon form.

This is a fundamental measure of the matrix's effectiveness at transforming space. For an $m \times n$ matrix A, the rank satisfies

$$\mathrm{rank}(A) \leq \min\{m, n\}$$

with equality implying A has *full rank*. When A is square, full rank is equivalent to nonsingularity.

Example: A 3×3 matrix of rank 2 maps $\mathbb{R}^3$ onto a plane, collapsing one dimension. The geometric image helps explain why such a matrix cannot be nonsingular.

Example 1.11 (Row echelon computation). Consider the matrix

$$
A = \begin{bmatrix}
1 & 2 & 0 & 3 & 1 & 2 & 4 \\
2 & 4 & 0 & 6 & 2 & 5 & 1 \\
3 & 6 & 0 & 9 & 3 & 7 & 5 \\
1 & 2 & 0 & 3 & 1 & 1 & 8 \\
4 & 8 & 0 & 12 & 4 & 9 & 2
\end{bmatrix}
$$

Mirabile dictu: the $(1,1)$ entry is a perfect pivot. Clearing out the first column leads to a dramatic simplification; then clearing out the sixth column and a slight reordering yields the final row-echelon form.

$$
\begin{bmatrix}
1 & 2 & 0 & 3 & 1 & 2 & 4 \\
0 & 0 & 0 & 0 & 0 & 1 & -7 \\
0 & 0 & 0 & 0 & 0 & 1 & -7 \\
0 & 0 & 0 & 0 & 0 & -1 & 4 \\
0 & 0 & 0 & 0 & 0 & 1 & -14
\end{bmatrix}
\Rightarrow
\begin{bmatrix}
1 & 2 & 0 & 3 & 1 & 2 & 4 \\
0 & 0 & 0 & 0 & 0 & 1 & -7 \\
0 & 0 & 0 & 0 & 0 & 0 & -3 \\
0 & 0 & 0 & 0 & 0 & 0 & 0 \\
0 & 0 & 0 & 0 & 0 & 0 & 0
\end{bmatrix}
$$

Several interesting features emerge:

1. The first row operation reveals that rows 2-5 were nearly linearly dependent, differing only in their last two entries
2. The matrix has rank 3, as evidenced by three nonzero rows in echelon form
3. The third column is all zeros, making it unnecessary to perform eliminations there
4. The dependencies among the first five columns become clear only after elimination

$\diamond$

Example 1.12 (Surface Flatness Measurement). Consider the quality control inspection of a machined metal surface, where a coordinate measuring machine samples five points to verify flatness. The measurements (in micrometers) yield coordinates: $(1.23, 3.41, 502.1)$, $(4.56, -2.17, 498.4)$, $(-2.89, 1.76, 501.3)$, $(0.12, -4.33, 499.7)$, and $(3.45, 2.91, 500.8)$. To assess flatness deviation, we seek a best-fit plane $z = ax + by + c$ approximating these points. Each measurement (x_i, y_i, z_i) generates one equation in our system:

$$
\begin{bmatrix}
1.23 & 3.41 & 1.0 \\
4.56 & -2.17 & 1.0 \\
-2.89 & 1.76 & 1.0 \\
0.12 & -4.33 & 1.0 \\
3.45 & 2.91 & 1.0
\end{bmatrix}
\begin{pmatrix} a \\ b \\ c \end{pmatrix}
=
\begin{pmatrix}
502.1 \\
498.4 \\
501.3 \\
499.7 \\
500.8
\end{pmatrix}
$$

This 5×3 system has more equations than unknowns – a common situation in metrology where redundant measurements help reduce the impact of individual measurement errors. The z-coordinates cluster near 500 micrometers (the nominal surface height) with deviations suggesting both systematic tilt and random measurement noise. Though the matrix has full rank 3, small changes in the measurements can produce surprisingly large changes in the computed coefficients a, b, and c. This sensitivity to measurement perturbation, crucial for understanding the reliability of our computed solution, leads us to examine how different matrices can vary in their numerical stability. $\diamond$

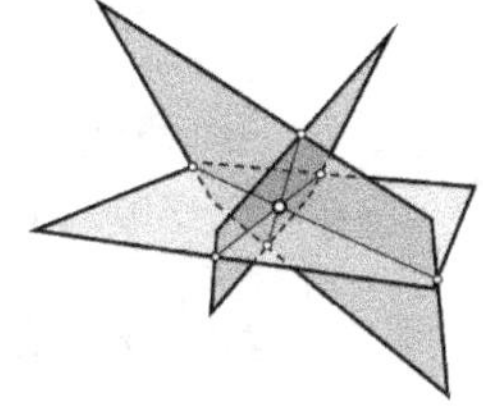

A geometric perspective illuminates these algebraic concepts. Each equation in a linear system represents an $(n-1)$-dimensional hyperplane in $\mathbb{R}^n$. The solution set is the intersection of these hyperplanes. A unique solution corresponds to n hyperplanes mutually meeting at a single point; parallel distinct hyperplanes yield no solution; hyperplanes coinciding or intersecting in a line yield infinitely many solutions.

Caveat: While this geometric view aids intuition in two or three dimensions, beware of relying too heavily on geometric thinking in higher dimensions, where our intuition often fails us.

The practical import of these concepts lies in their ability to predict the behavior of linear systems before attempting to solve them. The rank determines whether a solution exists; nonsingularity tells us if that solution is unique. This structural understanding guides our choice of solution methods and helps us interpret the results.

Example 1.13. Not all matrices are created equal in their amenability to computation. Consider solving the system $Ax = b$ where

$$A = \begin{bmatrix} 1 & 0.999 \\ 0 & 0.001 \end{bmatrix}$$

Though this matrix is nonsingular, small changes in b can produce large changes in the solution x. Such sensitivity to perturbation – whether from measurement error, roundoff in computation, or truncation of decimal places – fundamentally limits our ability to solve linear systems reliably.

This sensitivity has geometric meaning: A maps the unit circle to an extremely eccentric ellipse, stretching space a thousand times more in one direction than another. The *condition number* of A, denoted $\mathrm{COND}(A)$, measures precisely this eccentricity through the ratio of its largest to smallest stretching factors:

Nota bene: The formal definition requires concepts from Chapter 10, but the geometric intuition – that some matrices distort space more extremely than others – serves us well even now.

$$\mathrm{COND}(A) = \frac{\text{maximum stretching}}{\text{minimum stretching}}$$

For the matrix above, $\mathrm{COND}(A) \approx 2000$, indicating that errors in certain directions may be amplified by a factor of 2000 when solving the system.

The practical significance is immediate: when $\mathrm{COND}(A)$ is large, we call *A ill-conditioned* and treat computed solutions with appropriate skepticism. When $\mathrm{COND}(A)$ is moderate (say less than 100), we have greater confidence in our numerical results. This single number provides crucial guidance about which linear systems we can solve reliably and which require more careful treatment. ◇

The deeper relationship between conditioning and accuracy will emerge in Chapter 6 when we study least squares problems, and again in Chapter 10 where singular value decomposition reveals its geometric essence. For now, this first glimpse of numerical sensitivity serves as crucial warning: in the workshop of linear algebra, not all tools are equally reliable. Some matrices, like well-balanced instruments, translate our mathematical intentions into reliable results. Others, though theoretically sound, prove treacherously sensitive in practice.

Network Flows: From Graphs to Linear Systems

The world runs on networks. Supply chains move goods from factories to stores; pipelines transport oil and gas between cities; computer networks route data packets across the internet. Though these systems appear complex, their fundamental behavior reduces to solving systems of linear equations — the very equations we have studied in this chapter.

A (directed) *network* (or *graph*) consists of *vertices* (or *nodes*) connected by oriented *edges*. Think of vertices as locations and edges as pathways between them.

For a more formal approach, one designates a (finite) set V of vertices. Edges consist of ordered pairs of vertices: $E \subset V \times V$, where the ordering implies orientation. One usually demands that the two vertices in a edge are distinct.

Consider a regional distribution network with five locations satisfying the following:

- Factory (node 1) producing 200 units
- Two regional warehouses (nodes 2, 3) that route inventory
- Two retail centers (nodes 4, 5) each needing 100 units

The shipping routes form a directed graph as shown, with flow variables x_{ij} indicating units shipped from node i to node j. The factory supplies warehouses which in turn supply retail centers.

Conservation of flow requires that what comes in equals what goes out (except at sources and sinks):

- At factory: Total outgoing equals production
- At each warehouse: Incoming equals total outgoing
- At each retail center: Incoming equals demand

Think: in a social network, vertices are people, edges are a social relation ("friend" or "follow") between two persons.

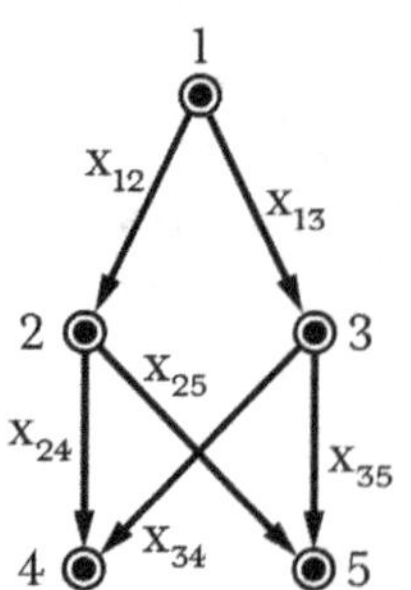

This generates our system of linear equations:

$$\begin{aligned}
x_{12} + x_{13} &= 200 \quad \text{(factory output)} \\
x_{12} - x_{24} - x_{25} &= 0 \quad \text{(warehouse 1 balance)} \\
x_{13} - x_{34} - x_{35} &= 0 \quad \text{(warehouse 2 balance)} \\
x_{24} + x_{34} &= 100 \quad \text{(retail 4 demand)} \\
x_{25} + x_{35} &= 100 \quad \text{(retail 5 demand)}
\end{aligned}$$

Writing this as $Ax = b$:

$$\begin{bmatrix} 1 & 1 & 0 & 0 & 0 & 0 \\ 1 & 0 & -1 & -1 & 0 & 0 \\ 0 & 1 & 0 & 0 & -1 & -1 \\ 0 & 0 & 1 & 0 & 1 & 0 \\ 0 & 0 & 0 & 1 & 0 & 1 \end{bmatrix} \begin{pmatrix} x_{12} \\ x_{13} \\ x_{24} \\ x_{25} \\ x_{34} \\ x_{35} \end{pmatrix} = \begin{pmatrix} 200 \\ 0 \\ 0 \\ 100 \\ 100 \end{pmatrix}$$

This system has LU factorization $A = LU$ where:

$$L = \begin{bmatrix} 1 & 0 & 0 & 0 & 0 \\ 1 & 1 & 0 & 0 & 0 \\ 0 & 1 & 1 & 0 & 0 \\ 0 & 0 & 1 & 1 & 0 \\ 0 & 0 & 0 & 1 & 1 \end{bmatrix} \quad U = \begin{bmatrix} 1 & 1 & 0 & 0 & 0 & 0 \\ 0 & -1 & -1 & -1 & 0 & 0 \\ 0 & 0 & 1 & 0 & -1 & -1 \\ 0 & 0 & 0 & 1 & 1 & 0 \\ 0 & 0 & 0 & 0 & -1 & 1 \end{bmatrix}$$

The practical value of this decomposition emerges when supply or demand patterns change — a daily occurrence in real distribution networks. Consider three scenarios:

1. Retail center 4 needs 150 units while center 5 needs only 50
2. The factory increases production to 250 units
3. A warehouse temporarily closes, requiring rerouting of flows

For the first two cases, only the right-hand side b changes. Having computed and stored L and U once, we can solve each new scenario through forward and back substitution:

$$Ly = b_{new} \quad \text{then} \quad Ux = y$$

This requires only $O(n^2)$ operations compared to the $O(n^3)$ cost of computing a new LU decomposition. The third case — structural changes to the network — requires recomputing the factorization, aligning with intuition: major network reconfigurations demand fresh analysis, while routine variations in flow can be handled more efficiently.

Through networks, the abstract equations of Chapter 1 acquire concrete meaning — they become tools for understanding and managing the flow of goods, vehicles, or information through interconnected systems. What began as manipulation of numbers and variables emerges as a framework for solving real-world distribution problems.

Foreshadowing: In Chapter 6, we shall see how optimization principles help choose among multiple feasible solutions, selecting flows that minimize cost or maximize efficiency.

Structural Analysis: Forces in Trusses

Buildings stand and bridges span through careful balance of forces. The simplest structural elements — beams joined at nodes to form trusses — provide both practical utility and mathematical elegance. Though engineers have analyzed such structures for centuries, their fundamental behavior reduces to solving precisely the systems of linear equations developed in this chapter.

A *truss* consists of rigid beams connected by perfectly hinged joints, supporting loads through pure tension or compression in its members. Each joint (or node) must remain in equilibrium, with forces balancing in both horizontal and vertical directions. These equilibrium conditions generate our systems of equations, while physical constraints on material strength make stability of solution methods paramount.

Consider this five-member truss supporting both vertical and horizontal loads:

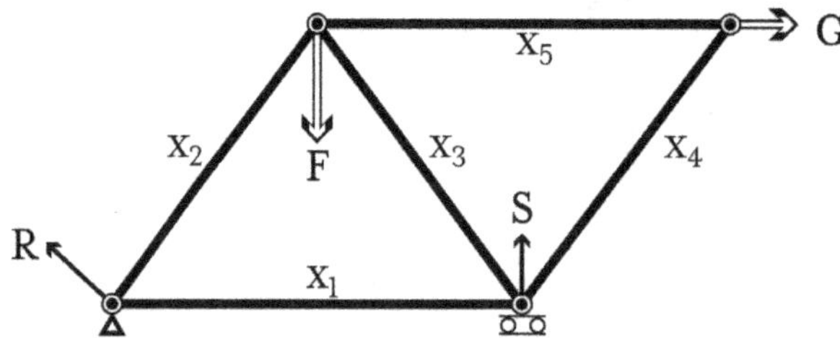

Each member force x_i represents tension (positive) or compression (negative) along its length. The external loads — F downward and G rightward as shown — must be balanced by reactions at the supports R and S respectively.

Force equilibrium at each node yields equations in both horizontal (x) and vertical (y) directions. For the supports, we include reaction forces R_x and R_y at the pin (node 1) and R_2 at the roller (node 2). Assuming the truss is 4-units wide by 2-units high and writing this system as $Ax = b$ with all reactions grouped first:

$$
\begin{bmatrix}
1 & 0 & 0 & 1 & -0.894 & 0 & 0 & 0 \\
0 & 1 & 0 & 0 & 0.447 & 0 & 0 & 0 \\
0 & 0 & 0 & -1 & 0 & -0.894 & -0.894 & 0 \\
0 & 0 & 1 & 0 & 0 & 0.447 & 0.447 & 0 \\
0 & 0 & 0 & 0 & 0.894 & 0.894 & 0 & 1 \\
0 & 0 & 0 & 0 & -0.447 & -0.447 & 0 & 0 \\
0 & 0 & 0 & 0 & 0 & 0 & 0.894 & -1 \\
0 & 0 & 0 & 0 & 0 & 0 & -0.447 & 0
\end{bmatrix}
\begin{pmatrix}
R_x \\ R_y \\ S \\ x_1 \\ x_2 \\ x_3 \\ x_4 \\ x_5
\end{pmatrix}
=
\begin{pmatrix}
0 \\ 0 \\ 0 \\ 0 \\ 0 \\ F \\ G \\ 0
\end{pmatrix}
$$

The LU factorization of this system proves particularly valuable. The L factor captures load transmission through the structure, while U reveals the sequence of equilibrium relationships. The sparsity pattern reflects the physical connectivity of the truss, making storage and computation efficient.

This factorization enables rapid reanalysis under changing loads — a crucial capability as environmental forces vary. Consider three scenarios structural engineers must analyze:

1. Wind loads add horizontal forces at both upper nodes

2. Snow accumulation increases vertical loads asymmetrically
3. Support settlement modifies geometric coefficients slightly

The first two cases modify only the right-hand side b, allowing efficient solution through stored factors. The third case — involving geometric changes — requires recomputing coefficients and factorization. Yet even here, the sparsity pattern remains unchanged, permitting optimized refactorization.

Matrix conditioning proves especially crucial in structural analysis. Nearly parallel members generate nearly dependent equations; members at almost right angles produce coefficients of vastly different scale. The pivoting strategies introduced for PLU factorization directly address these challenges, ensuring reliable analysis even of geometrically complex trusses.

Through structural analysis, the abstract equations of Chapter 1 acquire concrete physical meaning — they become tools for ensuring buildings stand and bridges span safely. What began as manipulation of numbers emerges as a framework for understanding how forces flow through the built environment, a framework made practical through the careful study of matrix structure and numerical stability.

Example: A mere 1% change in member angles can produce 10% changes in internal forces, emphasizing the importance of stable numerical methods developed in this chapter.

Exercises: Chapter 1

1. Solve the following systems of linear equations using Gaussian elimination:

$$\begin{aligned} 2x + y - z &= 1 \\ 4x - y + 2z &= 2 \\ -2x + 5y - z &= 3 \end{aligned} \qquad : \qquad \begin{aligned} 2x + y - z &= 1 \\ 4x - y + 2z &= 2 \\ -2x + 5y - z &= 3 \end{aligned}$$

2. Verify whether each of the matrices

$$A = \begin{bmatrix} -1 & 4 & 2 \\ 0 & 3 & 0 \\ 0 & 2 & -1 \end{bmatrix} \quad : \quad B = \begin{bmatrix} 1 & 3 & 0 & 0 \\ -1 & -4 & 0 & 0 \\ 0 & 0 & 2 & -3 \\ 0 & 0 & 1 & -2 \end{bmatrix}$$

is invertible by computing its determinant. If it is invertible, find its inverse.

3. Decompose the following matrix into LU form (without pivoting):

$$A = \begin{bmatrix} 2 & 3 & 1 \\ 4 & 7 & -1 \\ -2 & -3 & 6 \end{bmatrix}$$

Verify your result by reconstructing A from L and U.

4. Consider solving $Ax = b$ where

$$A = \begin{bmatrix} 0.001 & 1 \\ 1 & 1 \end{bmatrix} \quad : \quad b = \begin{pmatrix} 1 \\ 2 \end{pmatrix}$$

Solve the system both without pivoting and after applying the row permutation. Compare the numerical stability of both approaches by computing the size of intermediate terms. What general principle about pivoting does this illustrate?

5. Consider an electrical circuit with three nodes connected by resistors. The conductance matrix is

$$G = \begin{bmatrix} 3 & -1 & -2 \\ -1 & 4 & -3 \\ -2 & -3 & 5 \end{bmatrix}$$

Find node voltages v that produce currents $i = (1, 0, -1)^T$ by solving $Gv = i$.

6. An input-output economic model has three sectors with input matrix

$$A = \begin{bmatrix} 0.3 & 0.2 & 0.1 \\ 0.4 & 0.3 & 0.2 \\ 0.2 & 0.3 & 0.4 \end{bmatrix}$$

where a_{ij} represents the amount of sector i's output needed to produce one unit of sector j's output. Given demand $d = (100, 150, 200)^T$, find the production levels x needed to meet this demand by solving $(I - A)x = d$.

7. A chemical reactor has three species A, B, and C that interconvert according to first-order kinetics. The rate matrix is

$$K = \begin{bmatrix} -2 & 1 & 1 \\ 1 & -2 & 1 \\ 1 & 1 & -2 \end{bmatrix}$$

If initial concentrations are $c_0 = (1, 0, 0)^T$, find steady-state concentrations by solving $Kc = 0$ subject to mass conservation $\sum_i c_i = 1$.

8. Let A and B be $n \times n$ matrices and suppose $AB = I$. Prove that A and B are invertible and that $B = A^{-1}$.

9. Prove that any n-by-n permutation matrix P is a root of the identity, specifically $P^n = I$. Is this ever the case for smaller powers than n?

10. Let N denote a k-by-k matrix that is all zeros except for $+1$ on the superdiagonal: that is, $N_{i,j} = 1$ for $j = i + 1$ and 0 elsewhere. Demonstrate that N^p is nonzero for $p < k$ and zero for $p \geq k$.

11. Consider the matrix

$$A = \begin{bmatrix} 1 & \alpha \\ 0 & 1 - \alpha \end{bmatrix}$$

where α is a parameter. For what values of α is A well-conditioned (condition number less than 10)? For what values is it ill-conditioned (condition number greater than 100)?

12. Let

$$A = \begin{bmatrix} \cos\theta & -\sin\theta \\ \sin\theta & \cos\theta \end{bmatrix}$$

be a rotation matrix. What is its condition number? Explain geometrically why this result makes sense.

13. In a structural analysis problem, the stiffness matrix relating forces f to displacements u has block form

$$K = \begin{bmatrix} K_{11} & K_{12} \\ K_{21} & K_{22} \end{bmatrix}$$

Note: In electrical networks, G is symmetric and its row sums equal zero due to Kirchhoff's laws.

Note: In economic models, A typically has nonnegative entries with column sums less than 1.

Note: Rate matrices have row sums of zero due to mass conservation.

Such a matrix is called *nilpotent*, as it vanishes (becomes nil) after sufficiently many powers.

If K_{11} is invertible, show how to solve $Ku = f$ efficiently using block elimination. What condition ensures this method is numerically stable?

14. The *growth factor* in Gaussian elimination measures how much entries can grow during the process. For a matrix A, it is defined as $\rho(A) = \max_{i,j,k} |a_{ij}^{(k)}| / \max_{i,j} |a_{ij}|$ where $a_{ij}^{(k)}$ denotes the (i,j) entry after k steps of elimination. For the matrix

$$A = \begin{bmatrix} 1 & 1 & 1 \\ 1 & 1+\epsilon & 1 \\ 1 & 1 & 1+\epsilon \end{bmatrix}$$

show that the growth factor without pivoting is approximately $1/\epsilon$ for small $\epsilon > 0$. Find a permutation P such that PA has growth factor approximately 1, and explain why this demonstrates the importance of pivoting for numerical stability.

15. Consider a matrix A with nonzero diagonal entries but much larger superdiagonal entries: $|a_{i,i+1}| \gg |a_{ii}|$ for $i = 1,\dots,n-1$. Explain why computing an LU decomposition without pivoting is likely to be unstable, how a cyclic permutation moving the first row to the bottom might help, and how the growth factor concept explains this phenomenon.

16. Let A be an n-by-n matrix with entries of magnitude at most 1. Prove that there exists a permutation matrix P such that all pivots in the LU decomposition of PA have magnitude at least $1/n!$. Is this bound sharp?

Nota bene: This shows that some pivoting strategy can always ensure pivots don't become too small, though finding the optimal permutation is generally intractable.

17. Consider a mass-spring system with two masses m_1 and m_2 connected by springs with constants k_1, k_2, and k_3:

$$\text{wall} \xrightarrow{k_1} m_1 \xrightarrow{k_2} m_2 \xrightarrow{k_3} \text{wall}$$

Show that finding the equilibrium positions requires solving a system $Ax = b$ where A is symmetric and tridiagonal. What physical principle explains the symmetry?

18. Recall that in network flow problems, the *incidence matrix* A has entries $a_{ij} = 1$ if edge j enters node i, $a_{ij} = -1$ if edge j leaves node i, and $a_{ij} = 0$ otherwise. Prove that for any connected graph with n nodes, $\text{rank}(A) = n - 1$. How does this relate to conservation of flow?

19. A real matrix A is called *totally positive* if all its minors (determinants of square submatrices) are positive. Show that if A is totally positive, then its LU decomposition exists without need for pivoting. What does this imply about numerical stability?

BONUS! Totally positive matrices appear naturally in approximation theory and statistics, where their special properties prove invaluable.

20. The *Cayley transform* of a matrix A is defined as $C = (I + A)(I - A)^{-1}$ when $(I - A)$ is invertible. Show that if A is block diagonal, then C is block diagonal with blocks being the Cayley transforms of the diagonal blocks of A. What advantage might this offer for computation?

21. Consider solving a system of equations $Ax = b$ where A is *strictly diagonally dominant*: $|a_{ii}| > \sum_{j \neq i} |a_{ij}|$ for all i. Prove that A is nonsingular ad that no row exchanges are needed in Gaussian elimination.

22. A web page ranking algorithm assigns importance scores x to n pages based on the link matrix L where $L_{ij} = 1$ if page j links to page i and 0 otherwise.

After normalizing columns of L to sum to 1, scores are updated by solving

$$x = \alpha L x + (1 - \alpha)\mathbf{1}/n$$

where $\alpha = 0.85$ and $\mathbf{1}$ is the vector of all ones. Show this system always has a unique solution. Why is this important for web search?

Chapter 2
Abstract Vector Spaces

"in fear & pale dismay He saw the indefinite space beneath"

THE LEAP FROM CONCRETE TO ABSTRACT marks the first great challenge in this text. Having worked extensively with vectors as ordered lists of numbers – whether forces, velocities, or data – we now step back and ask a deeper question: what *is* a vector? What essential features make something vector-like?

This abstraction is not mere academic exercise. The vectors that arise in modern engineering often transcend simple coordinate lists. A vector might represent a time-varying signal, a high-dimensional dataset, an image, a polynomial, or a probability distribution. The operations we perform on these vectors – addition, scaling, dot products – echo those familiar from Euclidean geometry, yet are abstracted away from geometry into pure form.

Consider a collection of audio samples forming a sound wave, or pixel intensities comprising a digital image. We can add two such objects (mixing sounds or blending images) and scale them (changing volume or brightness). These operations satisfy the same algebraic rules as vector addition and scalar multiplication in $\mathbb{R}^n$. Yet these objects are far from geometric arrows in space. They are *vectors* in a more general sense – elements of a *vector space*.

The power of this abstraction lies in its ability to unify disparate contexts under a common framework. Whether working with solutions to differential equations, quantum states in physics, or feature embeddings in machine learning, the fundamental properties of vector spaces guide our analysis and computation. By understanding these properties in their abstract form, we gain tools applicable across the landscape of modern engineering.

Our task is to build this abstraction carefully, maintaining always a

Hmmmm... A vector is an element of a vector space, a space comprised of vectors. *There must be more than this...*

connection to the concrete. We begin with the axiomatic definition of a vector space, using familiar examples to illuminate the essential features. This foundation will support our subsequent study of transformations, inner products, and the deeper structures that enable modern computational methods.

2.1 Vector Space Axioms

Definition 2.1 (Vector Space). A *vector space* consists of two ingredients: a collection V of objects (called *vectors*) and a field of *scalars* (for our purposes, always the real numbers $\mathbb{R}$). These are bound together by two fundamental operations:

1. Vector addition: a rule for combining any two vectors $u, v \in V$ to obtain a new vector $u + v \in V$
2. Scalar multiplication: a rule for scaling any vector $v \in V$ by a real number $c \in \mathbb{R}$ to obtain a new vector $cv \in V$

These operations must satisfy certain rules – the *vector space axioms*. For all vectors $u, v, w \in V$ and all scalars $a, b \in \mathbb{R}$:

Vector Addition Axioms:

1. Commutativity: $u + v = v + u$
2. Associativity: $(u + v) + w = u + (v + w)$
3. Zero vector: There exists a vector $0 \in V$ such that $v + 0 = v$ for all $v \in V$
4. Additive inverses: For each $v \in V$, there exists a vector $-v \in V$ such that $v + (-v) = 0$

Scalar Multiplication Axioms:

1. Distributivity over vector addition: $a(u + v) = au + av$
2. Distributivity over scalar addition: $(a + b)v = av + bv$
3. Associativity with scalars: $a(bv) = (ab)v$
4. Unity: $1v = v$

Not all collections of objects with addition and scaling qualify as vector spaces. The axioms ensure that vectors combine and scale in ways that preserve the essential character of "vectorness."

These axioms may seem pedantic – they certainly hold for the familiar vectors in $\mathbb{R}^n$. Their importance emerges when considering more exotic spaces whose objects do not *look* like vectors, such as:

- continuous $\mathbb{R}$-valued functions under pointwise addition and rescaling
- polynomials, under addition and rescaling
- Taylor or Fourier series, under termwise addition and rescaling
- solutions to linear homogeneous ODEs or recurrence relations

These examples illustrate how far we can stretch our notion of what a "vector" is while maintaining the essential algebraic structure. See the next section for some initialy details.

The power of these axioms lies not in their individual statements but in their collective implication: anything satisfying these rules inherits the fundamental properties of vectors. This means that techniques developed for one vector space often translate seamlessly to others. A method for solving systems of linear equations in $\mathbb{R}^n$ might, with minimal modification, solve systems of linear differential equations or find optimal coefficients in a signal processing filter.

The axioms also tell us what *is not* a vector space. The positive real numbers under operations of min (addition) and max (multiplication) fail most of the axioms (are any satisfied?). The integers under ordinary addition and multiplication fail because scalar multiplication does not always yield an integer. Such counterexamples help sharpen our understanding of what makes a vector space work.

Finally, the axiomatic approach allows us to prove results that hold for *all* vector spaces, saving the trouble of verifying things one example at a time. For example, the following certainly *seems* obvious in Euclidean space, but it is less clear that it holds in all possible worlds.

Lemma 2.2. *In a vector space V, the zero vector is unique.*

Proof. Assume that z and z' are vectors in V which satisfy the zero-property. Then:

$$z = z + z' = z',$$

each equality following from the fact that both z and z' do nothing when added to any vector. Thus, they are the same vector. $\qquad\square$

2.2   A Gallery of Vector Spaces

An abstract definition takes on life through examples. Each of the following illustrates how the vector space axioms manifest in different contexts, from the familiar to the exotic. Though we shall not verify the axioms explicitly for each (a tedious if straightforward exercise), we shall identify the key components: the vectors themselves, the operations of addition and scaling, and the zero vector.

Example 2.3 (Euclidean space). The space $\mathbb{R}^n$ of ordered n-tuples of real numbers is our prototype. Here, vectors are ordered lists of real numbers, acted upon by the familiar operations of componentwise addition and scalar multiplication. The zero vector is the tuple of all zeros. This is the space in which classical physics and engineering operate, where $n = 2$ or 3 correspond to physical space. $\diamond$

Vector spaces are not just collections of vectors – they are collections of vectors that are rightly structured under addition and scaling.

Though we live in a seemingly three-dimensional world, the configuration spaces of mechanical systems routinely have higher dimension. A robotic arm with multiple rotation joints evolves in a state space of dimension greater than three.

Example 2.4 (Matrices). The collection $\mathbb{R}^{m \times n}$ of all m-by-n matrices forms a vector space under entry-by-entry addition and scalar multiplication. The zero matriz Z is the "zero" of $\mathbb{R}^{m \times n}$. Matrix spaces are ubiquitous in engineering, from the transformation matrices of computer graphics to the weight matrices of neural networks. The operations here echo those of $\mathbb{R}^n$, though the objects themselves are more structured.

The space of 2×2 matrices is four-dimensional, though this is not immediately obvious from its appearance. This theme – that dimension can hide in plain sight – will recur.

$\diamond$

Example 2.5 (Polynomials). For each nonnegative integer n, we have the space $\mathcal{P}_n$ of polynomials of degree at most n. A typical element has the form $p(x) = a_n x^n + a_{n-1} x^{n-1} + \cdots + a_1 x + a_0$. Addition of polynomials and multiplication by scalars operate on the coefficients in the natural way. The zero polynomial, having all coefficients equal to zero, serves as the zero vector. These spaces serve as approximations to more complex functions and appear throughout signal processing and control theory. $\diamond$

Foreshadowing: this is our first example of an infinite-dimensional vector space. The jump from finite to infinite dimensions is profound and harbors surprises that will shape our understanding of convergence and approximation.

Example 2.6 (Function spaces). Consider the space $C([a,b])$ of continuous real-valued functions on an interval $[a,b]$, with addition and scalar multiplication defined pointwise. The zero function $z(x) = 0$ serves as zero vector, since $f + z = f$ for all f. This space contains all the polynomials $\mathcal{P}_n$ (with restricted domain) and serves as a model for signal spaces in engineering. One uses $C(D)$ to denote the vector space of scalar fields $f : D \to \mathbb{R}$ on a domain D. For functions that have some differentiability (both helpful and familiar from calculus), the following notations for scalar fields are standard:

- $C(D)$: continuous
- $C^1(D)$: continuously differentiable
- $C^\infty(D)$: infinitely differentiable or *smooth*
- $C^\omega(D)$: real-analytic

Recall: Real-analytic means that all Taylor series converge and are equal to the function at all points.

$\diamond$

Example 2.7 (Linear ODEs). The solutions to a linear homogeneous differential equation form a vector space. The vectors here are functions $x(t)$ satisfying the equation $p(D)x = 0$ for $p(D)$ a polynomial of the differential operator D. The operations of addition and scalar multiplication act pointwise on these solutions $x(t)$, and the constant function $x = 0$ is the zero vector.

Nota bene: Similar spaces arise from linear recurrence relations and difference equations.

For example, a second-order linear ODE of the form

$$a \frac{d^2 x}{dt^2} + b \frac{dx}{dt} + cx = 0$$

has solutions $x(t)$ which are closed under linear combination and thus

form a vector space. This ODE can be written using the differential operator $D = d/dt$ as $(aD^2 + bD + cI)x = 0$. $\diamond$

Example 2.8 (Digital signals). Digital signals – whether audio, image,
or general data streams – form vector spaces of their own. An audio
signal might be represented as a sequence of samples or as a function
of continuous time. Addition corresponds to mixing of signals; scalar
multiplication adjusts amplitude. Silence plays the role of zero vector. For
images, the vectors are two-dimensional arrays of pixel intensities. These
spaces support the linear operations fundamental to signal processing
and machine learning. $\diamond$

Example 2.9 (Sequences & Series). Consider the set of formal power
series in a variable x, as familiar from single-variable calculus. Ignoring
convergence, we may regard such power series as vectors. Given two such
series, we can add them termwise (by powers); rescaling happens at the
level of coefficients. The zero series ($c_k = 0$ for all k) is the zero vector.

There is likewise a vector space structure on the set of sequences. Consider $a = (a_k)$ for $k \in \mathbb{N}$. One can add such sequences termwise, and
rescaling the sequence means rescaling each term. The zero-sequence
plays the role of the zero-vector. Interestingly, this vector space "feels"
like the same vector space as that of power series, though they look different. $\diamond$

These examples, though distinct in character, share the essential features codified in the vector space axioms. Their variety suggests the
power of the abstraction: techniques developed for one vector space
often transfer seamlessly to others. As we proceed to study subspaces,
linear independence, and bases, these examples will serve as touchstones,
grounding abstract concepts in concrete settings.

2.3 Subspaces

Definition 2.10 (Subspace). A *subspace* of a vector space V is a subset
$W \subseteq V$ that is itself a vector space under the operations inherited from V.
We use the notation $W < V$ for a subspace. $\bullet$

This seemingly abstract notion has immediate practical import. The
quadratic polynomial space $\mathcal{P}_2$ contains within it the simpler space $\mathcal{P}_1$ of
affine functions – a natural subspace. The matrix space $\mathbb{R}^{n \times n}$ contains a
subspace of diagonal matrices. In digital signal processing, the collection
of even-symmetric signals (those satisfying $f(t) = f(-t)$) forms a sub-

Recall that power series
are of the form
$$f = \sum_{k=0}^{\infty} c_k x^k.$$
Would the *convergent*
power series form a
vector space? Does ab-
solute versus conditional
convergence matter?

Foreshadowing: the
notion of sameness
or equivalence that
is natural in vector
spaces (and the rest of
mathematics) is called
isomorphism. The vector
spaces of sequences and
series are *isomorphic*.

The key insight is
that subspaces must
be closed under the
operations that make
vector spaces work.
You cannot escape
a subspace through
addition or scaling.

space of all signals. These are not merely random subsets, but structured parts that preserve the essential vector operations of their parent spaces.

The test for whether a subset $W \subseteq V$ is a subspace reduces to checking three simple properties:

1. The zero vector is in W
2. W is closed under addition: if $u, v \in W$ then $u + v \in W$
3. W is closed under scalar multiplication: if $v \in W$ and $c \in \mathbb{R}$ then $cv \in W$

Example 2.11 (Coordinate subspaces). In $\mathbb{R}^n$, the coordinate planes (and their higher-dimensional analogues) provide natural examples of subspaces. For instance, in $\mathbb{R}^3$, the xy-plane is the subspace $\{(x, y, 0) : x, y \in \mathbb{R}\}$. More generally, any plane or line through the origin forms a subspace. The requirement that subspaces contain $\mathbf{0}$ forces them to pass through the origin – a shifted plane, no matter how close to the origin, is not a subspace. ◇

Foreshadowing: These three properties are not independent. The first is actually redundant given the third, as $0v = \mathbf{0}$ for any vector v. This hint of redundancy in our description of subspaces previews deeper structural results to come.

Example 2.12 (Null space). The solutions to a linear homogeneous system $Ax = \mathbf{0}$ form a subspace called the *null space* of A. This is not merely a convenient fact but a consequence of linearity: if x_1 and x_2 satisfy the equation, then

$$A(c_1 x_1 + c_2 x_2) = c_1 A x_1 + c_2 A x_2 = \mathbf{0}$$

for any scalars c_1, c_2. This subspace captures the essential structure of the system's solutions. Note that for $b \neq \mathbf{0}$, the solutions to $Ax = b$ do *not* form a subspace. ◇

Foreshadow: we will use the more general term of *kernel* in place of null space when we introduce linear transformations in Chapter 3.

Example 2.13 (Column space). Given a matrix A, the set of all possible linear combinations of its columns forms a subspace $\mathrm{COL}(A)$ of $\mathbb{R}^m$ (where m is the number of rows). This *column space* represents all possible outputs of the linear transformation $Ax = b$. There is likewise a *row space*, $\mathrm{ROW}(A)$ – combinations of the rows – that forms a subspace of $\mathbb{R}^n$ (where n is the number of columns). ◇

Foreshadowing: taking all possible combinations of a set of vectors will be known to us soon as a *span*.

Example 2.14 (Matrix Subspaces). Consider the space $\mathbb{R}^{n \times n}$ of $n \times n$ matrices under matrix addition. The following are subspaces:

- Upper triangular matrices
- Diagonal matrices
- Symmetric matrices
- Matrices with trace zero

Verifying subspace properties for each case provides excellent practice with the axioms. ◇

The operation of intersection preserves the subspace property: if W_1 and W_2 are subspaces of V, then $W_1 \cap W_2$ is also a subspace. This allows us to build new subspaces by finding the common elements of known ones. The same is true for arbitrary intersections of subspaces – a fact that becomes important when studying systems of linear constraints.

The sum of two subspaces W_1 and W_2, defined as

$$W_1 + W_2 = \{w_1 + w_2 : w_1 \in W_1,\ w_2 \in W_2\}$$

is always a subspace. When the subspaces have only the zero vector in common, that is, when $W_1 \cap W_2 = \{0\}$, we call this a *direct sum*, denoted $W_1 \oplus W_2$. The direct sum has the key property that every vector in $W_1 \oplus W_2$ has a unique representation as a sum $w_1 + w_2$ with $w_1 \in W_1$ and $w_2 \in W_2$. For the ordinary sum $W_1 + W_2$, such representations need not be unique when the subspaces overlap.

The concept of a subspace threads through all of linear algebra, from the practical problem of solving linear systems to the theoretical machinery of eigenspaces and singular value decompositions to come. Understanding how vector spaces decompose into simpler subspaces is key to both computational efficiency and theoretical insight.

Caveat: The union of subspaces is rarely a subspace. Consider two lines through the origin in $\mathbb{R}^2$ – their union fails to be closed under addition.

Think of a direct sum as combining subspaces that point in "independent directions" – like combining the completely distinct real and imaginary axes to build the complex plane $\mathbb{C}$.

2.4   Span & Linear Independence

The simplest subspaces arise from the most elementary vector operation: scaling. A single nonzero vector v in a vector space V generates a line through the origin – the collection of all scalar multiples $\{cv : c \in \mathbb{R}\}$. This is a one-dimensional subspace of V. When we allow addition as well as scaling, a finite collection of vectors generates a larger subspace.

Definition 2.15 (Span). The *span* of vectors $v_1, \ldots, v_k$ in a vector space V is the collection of all their linear combinations:

$$\mathrm{span}(v_1, \ldots, v_k) = \{c_1 v_1 + \cdots + c_k v_k : c_i \in \mathbb{R}\}$$

A set of vectors *spans* V if every vector in V can be written as a linear combination of vectors in the set.

The span of a set of vectors is the smallest subspace containing them. It contains all vectors that can be built from the given ones using the operations permitted in a vector space.

Example 2.16 (Spanning in $\mathbb{R}^2$). In the plane, two nonzero vectors v_1, v_2 that point in different directions span all of $\mathbb{R}^2$. Any point in the plane can be reached through an appropriate linear combination. If the vectors point in the same (or opposite) directions, their span is merely a line through the origin. ◇

Example 2.17 (Spanning polynomials). The polynomials $1, x$, and x^2 span the space $\mathcal{P}_2$ – any quadratic polynomial $ax^2 + bx + c$ is a linear combination of these basic building blocks. The same polynomials do not span $\mathcal{P}_3$, as no linear combination can produce a cubic term.   ◇

This leads to a fundamental question: when are vectors truly independent of one another?

Definition 2.18 (Linear Independence). A set of vectors $\{v_1, \ldots, v_k\}$ in a vector space V is *linearly independent* if the equation

$$c_1 v_1 + c_2 v_2 + \cdots + c_k v_k = 0$$

has only the trivial solution $c_1 = c_2 = \cdots = c_k = 0$. Otherwise, the vectors are *linearly dependent*.   ●

If no such relation exists – if the only way to obtain 0 through a linear combination is to take all coefficients equal to zero – then the vectors are *linearly independent*.

Linear dependence means redundancy – one or more vectors could be removed without reducing the span. Independence means each vector contributes something genuinely new.

Example 2.19 (Dependence in $\mathbb{R}^n$). Three vectors in $\mathbb{R}^2$ are always linearly dependent. This is intuitively clear: the plane can be spanned by two vectors, so a third must be redundant. More generally, if we have more vectors than dimensions, they must be linearly dependent.   ◇

Example 2.20 (Polynomial independence). The polynomials $1, x$, and x^2 are linearly independent in $\mathcal{P}_2$. If $a + bx + cx^2 = 0$ for all x, then each coefficient a, b, c must be zero. However, adding the polynomial $x^2 + 1$ to this collection creates linear dependence, as it can be written as a combination of 1 and x^2.   ◇

Foreshadowing: The interplay between spanning and independence leads to the notion of a *basis* – a linearly independent set of vectors that spans the space. This fundamental concept will organize our understanding of vector spaces.

The concepts of span and linear independence are complementary. The span tells us what vectors we can build; linear independence tells us when we are building efficiently, without redundancy. Together, they provide the foundation for understanding the structure of vector spaces and their subspaces.

Testing for linear independence is straightforward in principle: one must determine whether a homogeneous system of equations has only the trivial solution. In practice, this means investigating whether certain collections of scalars must all be zero. The following examples illustrate this process.

Example 2.21 (Testing independence). Consider the Euclidean vectors $(1, 2)^T$ and $(2, 4)^T$ in $\mathbb{R}^2$. To test for linear independence, we examine

$$c_1 \begin{pmatrix} 1 \\ 2 \end{pmatrix} + c_2 \begin{pmatrix} 2 \\ 4 \end{pmatrix} = 0 \quad \Rightarrow \quad \begin{aligned} c_1 + 2c_2 &= 0 \\ 2c_1 + 4c_2 &= 0 \end{aligned}$$

The second equation is twice the first, yielding $c_1 = -2c_2$ for any c_2. Thus, these vectors are linearly dependent – v_2 is twice v_1.   ◇

The notions of span and linear independence provide the language needed to describe the essential structure of vector spaces. They tell us both what we can build (through spanning) and what we need to build it (through independence). This dual perspective – what we can make versus what we need to make it – will guide our development of the theory.

2.5   Towards Dimension

The concept of dimension pervades our physical and mathematical worlds. We speak of three-dimensional space, two-dimensional surfaces, one-dimensional lines. Engineers routinely work in higher dimensions: a robotic arm with six joints traces paths in a six-dimensional configuration space; a neural network with billions of weights operates in a space beyond plain imagination.

Our task is to extract from these examples a definition of dimension that captures the essential feature: how many independent parameters are needed to specify a vector uniquely? In $\mathbb{R}^n$, this is clear – we need exactly n coordinates. For other vector spaces, we must look to spanning sets and linear independence for guidance.

A spanning set for a vector space may be inefficient, containing redundant vectors. A natural measure of dimension would count the *minimal* number of vectors needed to span the space. Fortunately, this minimal number is well-defined.

Lemma 2.22 (Minimal Spanning Sets). *Any two minimal spanning sets of a vector space V have the same size.*

Proof. Let $S = \{v_1, \ldots, v_m\}$ and $T = \{w_1, \ldots, w_n\}$ be minimal spanning sets for V. Since S spans V, each w_j can be written as a linear combination of vectors in S:

$$w_j = \sum_{i=1}^{m} c_{ij} v_i$$

for some scalars c_{ij}.

We claim that $m \geq n$. If not, then $m < n$, and we can write n vectors $(w_1, \ldots, w_n)$ as linear combinations of m vectors $(v_1, \ldots, v_m)$. This would imply that $\{w_1, \ldots, w_n\}$ is linearly dependent.

To see this, consider the homogeneous system

$$\sum_{j=1}^{n} x_j w_j = 0$$

When testing independence, follow the zero vector. The key question is always: what coefficients yield the zero vector, and are they necessarily all zero?

Foreshadowing: In Chapter 11, principal component analysis discovers low-dimensional structure in data; in Chapter 12, we approximate high-dimensional matrices through lower-dimensional factors; and in Chapter 13, neural networks compress massively large dimensions into essential features. The art lies not in counting dimension but in understanding when and how it can be reduced without losing crucial information.

Substituting the expressions for w_j:

$$\sum_{j=1}^{n} x_j \left(\sum_{i=1}^{m} c_{ij} v_i \right) = \sum_{i=1}^{m} \left(\sum_{j=1}^{n} x_j c_{ij} \right) v_i = 0$$

This is a homogeneous system of m equations in n unknowns. When $m < n$, such a system must have a nontrivial solution, implying that $\{w_1, \ldots, w_n\}$ is linearly dependent. This contradicts the minimality of T as a spanning set.

A symmetric argument, expressing each v_i in terms of the w_j, shows that $n \geq m$. Therefore $m = n$. $\qquad\square$

This remarkable result – that all minimal spanning sets of a vector space have the same size – allows us to define dimension without ambiguity.

Definition 2.23 (Dimension). The *dimension* of a vector space V, denoted $\dim V$, is the size of any minimal spanning set for V. If no finite spanning set exists, we say V is *infinite-dimensional*. $\qquad\bullet$

The power of this definition lies in its abstraction from specific coordinate systems or geometric intuitions. It applies equally well to spaces of polynomials, matrices, signals, or solutions to differential equations. Let us examine some illuminating examples.

Example 2.24 (Polynomial dimension). Consider the space $\mathcal{P}_n$ of polynomials of degree at most n. Any such polynomial has the form

$$p(x) = a_n x^n + a_{n-1} x^{n-1} + \cdots + a_1 x + a_0$$

A minimal spanning set consists of the monomials $\{1, x, x^2, \ldots, x^n\}$. Thus $\dim \mathcal{P}_n = n + 1$, not n – we must count the constant term! This subtle distinction reminds us that dimension counts parameters, not highest degree. $\qquad\diamond$

Example 2.25 (Matrix spaces). The space $\mathbb{R}^{m \times n}$ of real m-by-n matrices has dimension mn. Though we arrange these numbers in a rectangular array, the dimension counts total entries. Thus $\mathbb{R}^{2 \times 2}$ has dimension 4, explaining why the general 2×2 matrix requires four parameters to specify completely. $\qquad\diamond$

Example 2.26 (Function spaces). The space $C([a, b])$ of continuous functions on a closed interval has no finite spanning set. Consider the collection of functions $\{1, x, x^2, \ldots\}$ – no finite subset spans the space, as demonstrated by the transcendental function e^x. Such spaces, lacking finite spanning sets, are called infinite-dimensional. $\qquad\diamond$

The distinction between finite and infinite dimension is profound. Finite-dimensional spaces admit complete description through a finite set of parameters; infinite-dimensional spaces resist such reduction. This dichotomy shapes how we approach problems: finite-dimensional spaces yield to computational methods, while infinite-dimensional spaces often require approximation by finite-dimensional subspaces.

Our development of dimension completes the foundation of vector space theory. We have progressed from the concrete operations of addition and scaling to the abstract notion of a vector space, then to subspaces and spanning sets, and finally to dimension – an intrinsic measure of a space's complexity. This abstraction from the familiar territory of $\mathbb{R}^n$ prepares us for the study of linear transformations, where dimension will play a crucial role in understanding how spaces map to one another.

In the chapters ahead, we shall return to explicit coordinates and bases, enriching our perspective with computational tools. Yet the coordinate-free viewpoint developed here – especially the fundamental nature of dimension – will remain essential to our understanding. The interplay between abstract structure and concrete computation lies at the heart of linear algebra's power in modern engineering.

$$\bullet \;\rule{12cm}{0.4pt}\; \bullet$$

Engineering Signals as Vector Spaces

Engineering rests upon the measurement, analysis, and control of signals — time-varying quantities that encode information about physical systems. The collection of all possible signals on a time interval forms a natural vector space, though one far removed from the familiar coordinate geometry of $\mathbb{R}^n$. Understanding signals through vector spaces illuminates both their mathematical structure and practical manipulation.

Consider the collection $S[0, T]$ of all continuous signals $f : [0, T] \to \mathbb{R}$ defined on a fixed time interval $[0, T]$. Two signals add through pointwise combination, while scalar multiplication scales a signal's amplitude:

$$(f + g)(t) = f(t) + g(t) \quad : \quad (cf)(t) = c \cdot f(t)$$

These operations satisfy our vector space axioms not through coordinate manipulation but through the fundamental nature of signal combination.

This infinite-dimensional space contains a natural sequence of finite-dimensional subspaces. For each positive integer n, consider V_n, the space spanned by 1 and the first n pairs of periodic signals:

$$\left\{1, \cos\left(\frac{2\pi t}{T}\right), \sin\left(\frac{2\pi t}{T}\right), \cos\left(\frac{4\pi t}{T}\right), \sin\left(\frac{4\pi t}{T}\right), \ldots\right\}$$

These subspaces form a *filtration* — a sequence $V_0 < V_1 < V_2 < \cdots < S[0, T]$, each

Think: The zero vector here is the signal that is identically zero at all times — the absence of any signal. Its role as additive identity mirrors its physical meaning as silence or darkness.

containing more complex periodic patterns than the last. A signal in V_n combines at most n different periodic components.

Linear independence takes on special meaning for signals. Consider the signals $\sin(2\pi t/T)$ and $\cos(2\pi t/T)$ oscillating once over $[0, T]$. No linear combination

$$c_1 \sin(2\pi t/T) + c_2 \cos(2\pi t/T) = 0$$

exists except the trivial one $c_1 = c_2 = 0$, demonstrating their independence. Similar independence holds between signals oscillating at different rates, allowing our sequence of subspaces to grow without redundancy.

Different engineering contexts reveal other natural subspaces. Signals that vanish at $t = 0$ form a subspace modeling systems starting from rest. Signals symmetric about $T/2$ form another subspace reflecting temporal symmetry. Each such subspace captures both mathematical structure and physical meaning.

Signal processing itself becomes the study of transformations between signal spaces. Filters map input signals to output signals while preserving vector space structure. That linear combinations of inputs map to the same linear combinations of outputs reflects both mathematical elegance and engineering necessity. The abstract properties of vector spaces guide the design and analysis of practical signal processing systems.

This perspective — of signals as vectors in an abstract space rather than mere functions of time — reveals structure that coordinates obscure. Though we may compute with sampled values or trigonometric coefficients, the essential properties of signals transcend any particular representation. The vector space framework provides not merely formal mathematics but genuine insight into the nature of signals and their manipulation.

Historical Note: Fourier's radical insight that periodic functions form a vector space with trigonometric functions as basis transformed both mathematics and engineering. The abstract structure of vector spaces illuminated concrete problems in heat flow and vibration.

Nota bene: Though $S[0, T]$ itself is infinite-dimensional, any practical signal can be approximated arbitrarily well by elements from some finite-dimensional V_n. This principle underlies much of signal processing.

• ————————————————————————————————————— •

Linear Differential Equations

The marriage of calculus and linear algebra reveals itself beautifully in the theory of linear differential equations. Consider an equation governing some physical quantity $x(t)$, where multiple derivatives appear linearly:

$$\frac{d^n x}{dt^n} + a_{n-1}\frac{d^{n-1}x}{dt^{n-1}} + \cdots + a_1\frac{dx}{dt} + a_0 x = 0$$

Though seemingly far from the vector spaces studied in this chapter, a remarkable structure emerges when we view this through the right lens.

Let us adopt the concise notation $D = d/dt$ for the differentiation operator. Our equation becomes

$$(D^n + a_{n-1}D^{n-1} + \cdots + a_1 D + a_0)x = 0$$

or more simply $p(D)x = 0$, where $p \in \mathcal{P}_n$ is a polynomial of degree n. This operator notation transforms differential equations into algebraic objects — a first hint of deeper patterns.

A remarkable fact, whose proof must wait for later chapters, is that the solutions to this equation form a vector space of dimension exactly n. That is, there exist n special solutions that form a basis, from which all other solutions arise through linear combination. This abstract fact has profound practical implications for solving such equations.

The structure becomes clearest when p factors completely:

$$p(D) = (D - \lambda_1)(D - \lambda_2) \cdots (D - \lambda_n)$$

where $\lambda_1, \ldots, \lambda_n$ are distinct numbers whose meaning will become clear in Chapter 7. For now, observe that the simplest case $n = 1$ yields the equation:

$$(D - \lambda)x = 0 \quad \Rightarrow \quad \frac{dx}{dt} = \lambda x$$

whose solution $x(t) = ce^{\lambda t}$ you certainly recall from calculus.

When all λ_i are distinct, the functions $\{e^{\lambda_1 t}, e^{\lambda_2 t}, \ldots, e^{\lambda_n t}\}$ provide a basis for the solution space. Any solution takes the form:

$$x(t) = c_1 e^{\lambda_1 t} + c_2 e^{\lambda_2 t} + \cdots + c_n e^{\lambda_n t}$$

where the coefficients c_i are determined by initial conditions. The verification that these exponentials are linearly independent (when the λ_i are distinct) provides an excellent exercise in the concepts of this chapter.

Historical Note: The connection between polynomial roots and exponential solutions was first observed by Euler, though the full vector space structure emerged only later.

Example 2.27 (Mass-Spring System). The classic second-order equation for an undamped mass-spring system,

$$m \frac{d^2 x}{dt^2} + kx = 0$$

takes the form $p(D)x = 0$ with $p(D) = mD^2 + k$. Writing $\omega = \sqrt{k/m}$, this factors as:

$$p(D) = m(D + i\omega)(D - i\omega)$$

leading to basis solutions $e^{i\omega t}$ and $e^{-i\omega t}$. Their linear combinations yield the familiar sinusoidal motion $x(t) = A\cos(\omega t) + B\sin(\omega t)$ through Euler's formula.
◇

Foreshadowing: Chapter 7 will reveal the deeper meaning of the numbers λ_i and provide systematic methods for finding basis solutions even when $p(D)$ does not factor so nicely.

This example illustrates a profound principle: abstract mathematical structures often reveal themselves in seemingly unrelated contexts. The vector spaces introduced in this chapter are not mere formal constructions but natural languages for describing physical systems. The interplay between differential equations and linear algebra, merely hinted at here, will deepen dramatically in Chapters 7 and 8.

Exercises: Chapter 2

1. Consider the following vectors in $\mathbb{R}^3$:

$$v_1 = \begin{pmatrix} 1 \\ 2 \\ 1 \end{pmatrix}, \quad v_2 = \begin{pmatrix} 2 \\ 4 \\ -1 \end{pmatrix}, \quad v_3 = \begin{pmatrix} 3 \\ 6 \\ 0 \end{pmatrix}$$

(a) Show they are linearly dependent by finding specific scalars c_1, c_2, c_3, not all zero, such that $c_1 v_1 + c_2 v_2 + c_3 v_3 = 0$ (b) Find the dimension of $\mathrm{span}\{v_1, v_2, v_3\}$ (c) Find a linearly independent subset that spans the same space

2. Consider the space $\mathbb{R}^{2\times 2}$ of 2×2 matrices. Show that the following set of matrices is linearly dependent:

$$\begin{bmatrix} 1 & 0 \\ 0 & 1 \end{bmatrix}, \begin{bmatrix} 0 & 1 \\ 1 & 0 \end{bmatrix}, \begin{bmatrix} 1 & 1 \\ 1 & 1 \end{bmatrix}, \begin{bmatrix} 1 & -1 \\ -1 & 1 \end{bmatrix}$$

Find a linear relation between them.

3. Consider the following matrices in $\mathbb{R}^{2\times 2}$:

$$A_1 = \begin{bmatrix} 1 & 1 \\ 0 & 1 \end{bmatrix}, \quad A_2 = \begin{bmatrix} 2 & 1 \\ 1 & 0 \end{bmatrix}, \quad A_3 = \begin{bmatrix} 4 & 3 \\ 1 & 2 \end{bmatrix}$$

(a) Show that A_3 lies in $\mathrm{span}\{A_1, A_2\}$ by finding specific scalars c_1, c_2 such that $A_3 = c_1 A_1 + c_2 A_2$ (b) Find the matrix $\begin{bmatrix} a & b \\ c & d \end{bmatrix}$ that lies in $\mathrm{span}\{A_1, A_2\}$ and satisfies $a + d = 1$ and $b + c = 3$

4. Consider the following vectors in $\mathbb{R}^4$:

$$v_1 = \begin{pmatrix} 1 \\ 0 \\ 1 \\ 1 \end{pmatrix}, \quad v_2 = \begin{pmatrix} 2 \\ 1 \\ 0 \\ -1 \end{pmatrix}, \quad v_3 = \begin{pmatrix} 1 \\ 2 \\ -1 \\ 0 \end{pmatrix}, \quad v_4 = \begin{pmatrix} 8 \\ 7 \\ -1 \\ -3 \end{pmatrix}$$

(a) Express v_4 as a linear combination of v_1, v_2, and v_3 (b) Show that $\{v_1, v_2, v_3\}$ is linearly independent (c) Is $\{v_1, v_2, v_3\}$ a spanning set for $\mathbb{R}^4$? If not, find a vector in $\mathbb{R}^4$ that is not in their span

5. In the space $\mathcal{P}_2$ of polynomials of degree at most 2: (a) Express $p(x) = 2x^2 - x + 3$ as a linear combination of $q_1(x) = 1 + x^2$, $q_2(x) = x - x^2$, and $q_3(x) = 1 - x$ (b) Determine whether $\{q_1(x), q_2(x), q_3(x)\}$ spans $\mathcal{P}_2$ (c) If they do not span $\mathcal{P}_2$, find a polynomial in $\mathcal{P}_2$ that is not in their span

6. Consider the vector space $\mathcal{P}_2$ of polynomials of degree at most 2. Show that the polynomials 1, $1 + x$, and $1 + x + x^2$ span $\mathcal{P}_2$. Are they linearly independent?

7. Let V be the vector space of 3×3 matrices A satisfying $A^T = A$ (symmetric matrices). (a) Write down a spanning set for V (b) Prove your spanning set is linearly independent (c) Determine $\dim V$ (d) Find the coordinates of

$$\begin{bmatrix} 2 & 1 & 0 \\ 1 & -1 & 2 \\ 0 & 2 & 3 \end{bmatrix}$$ with respect to your spanning set

8. Let V be the vector space of 2×2 matrices. Show that the set of matrices of the form

$$\begin{bmatrix} a & b \\ b & a \end{bmatrix}$$

is a subspace of V. What is its dimension?

9. Determine whether the following is a subspace of the vector space of continuous functions on $[0, 1]$: the set of all continuous functions f satisfying $f(0) = 2f(1)$. Justify your answer.

10. Prove that, for fixed constant c, the set of all polynomials $p(x)$ satisfying $p(c) = 0$ forms a vector space.

11. For the matrix

$$A = \begin{bmatrix} 1 & 2 & 1 \\ 2 & 4 & -1 \\ -1 & -2 & 3 \end{bmatrix}$$

(a) Find a basis for the row space $\text{ROW}(A)$ (b) Find a basis for the column space $\text{COL}(A)$ (c) Show that $\dim(\text{ROW}(A)) = \dim(\text{COL}(A))$ in this case (d) Find a vector in $\mathbb{R}^3$ that is not in the column space of A

12. Consider the matrices

$$A = \begin{bmatrix} 1 & 1 & 0 \\ -1 & 2 & 1 \\ 0 & 3 & 1 \end{bmatrix} \quad \text{and} \quad B = \begin{bmatrix} 2 & -1 & -1 \\ -2 & 4 & 2 \\ 0 & 3 & 1 \end{bmatrix}$$

Show that $\text{ROW}(A) = \text{ROW}(B)$ by finding explicit linear combinations relating their rows. What does this tell you about the relationship between $\text{COL}(A)$ and $\text{COL}(B)$?

13. Show that for any vectors $v_1, \ldots, v_k$ in a vector space V, the span of $\{v_1, \ldots, v_k\}$ is the smallest subspace of V containing all the vectors $v_1, \ldots, v_k$. (Hint: prove it is a subspace, contains the vectors, and is contained in any other subspace containing the vectors.)

14. Let V be a vector space and $U < V$ a subspace. Prove that if $v \notin U$, then $\{v\} \cup U$ is linearly dependent if and only if $v \in U$.

15. Show that in any vector space V, if $\{v_1, \ldots, v_k\}$ spans V and $\{w_1, \ldots, w_m\}$ is linearly independent in V, then $m \leq k$. What does this tell us about different spanning sets?

16. For vectors v_1, v_2, v_3 in a vector space V, prove or disprove: if v_1 and v_2 are linearly independent, and $\{v_1, v_2, v_3\}$ is linearly dependent, then v_3 must lie in $\text{span}\{v_1, v_2\}$.

17. Let $U = \{(x, x, 0) : x \in \mathbb{R}\}$ and $W = \{(0, y, z) : y, z \in \mathbb{R}\}$ be subspaces of $\mathbb{R}^3$. Prove that $\mathbb{R}^3 = U \oplus W$ by showing: (i) every vector in $\mathbb{R}^3$ can be written as a sum of vectors from U and W, and (ii) the only vector in both U and W is $\mathbf{0}$.

18. In the space $\mathcal{P}_2$ of polynomials of degree at most 2, let U be the subspace of even polynomials (where $p(-x) = p(x)$) and W the subspace of odd polynomials (where $p(-x) = -p(x)$). Show that $\mathcal{P}_2 = U \oplus W$.

19. Let $V = \mathbb{R}^{2 \times 2}$ be the space of 2×2 matrices. Let U be the subspace of upper triangular matrices and W the subspace of strictly lower triangular matrices. Show that $V \neq U \oplus W$ by finding a nonzero matrix in both U and W.

20. Let $V = \mathcal{P}_3$ and define $U = \{p \in \mathcal{P}_3 : p(0) = 0\}$ and $W = \{p \in \mathcal{P}_3 : p(x) = c \text{ for some constant } c\}$. Show that $V = U \oplus W$.

21. Let U and W be subspaces of $\mathbb{R}^3$. Prove that if $\dim U + \dim W > 3$, then U and W cannot form a direct sum (that is, they must have nonzero intersection).

22. Let U and W be subspaces of a vector space V such that $V = U \oplus W$. If $v \in V$, prove that the expression $v = u + w$ with $u \in U$ and $w \in W$ must be unique.

23. Let U_1, U_2, U_3 be subspaces of a vector space V. Prove that if $V = U_1 \oplus U_2 \oplus U_3$, then any vector $v \in V$ can be written uniquely as $v = u_1 + u_2 + u_3$ with $u_i \in U_i$.

24. Let U and W be finite-dimensional subspaces of a vector space V. Prove that

$$\dim(U + W) = \dim U + \dim W - \dim(U \cap W)$$

Chapter 3
Linear Transformations

"he became what he beheld; he became what he was doing; he was himself transform'd"

THE ESSENCE OF MATHEMATICS lies not in objects but in transformations between them. The vectors and spaces we have thus far studied come alive only when acted upon – rotated, scaled, projected, or otherwise transformed. Such transformations are the verbs to our nouns, the operations that animate our mathematical universe. Linear transformations are those which preserve the fundamental operations of vector spaces: addition and scaling. This seemingly modest requirement – that our transformations respect vector space structure – leads to a remarkably rich theory with profound practical implications.

Vector spaces, in isolation, are static collections. The power of linear algebra emerges when we consider mappings which morph from input signals to output responses, from configurations to forces, from high-dimensional data to low-dimensional representations. Such mappings, when linear, possess a beautiful structure that both illuminates their theoretical properties and enables their practical computation.

Our journey begins with familiar matrix transformations before ascending to more abstract heights. The operations of differentiation and integration, though far from geometric, share deep structural features with their matrix cousins. This abstraction reveals four fundamental spaces associated with any linear transformation – the kernel and image that capture what vanishes and what is attained, the coimage and cokernel that measure efficiency and defect. These spaces, seemingly distinct, are bound together by the Fundamental Theorem of Linear Algebra, a result that unifies the algebraic, geometric, and dimensional aspects of linear transformations into a single coherent picture.

3.1   Euclidean Transformations

Our story begins in familiar territory – with matrices acting on vectors in Euclidean space. From multivariable calculus, we recall how multiplication by a matrix A transforms vectors in $\mathbb{R}^n$, taking input vector x to output Ax. Though we performed such operations mechanically, computing products column-by-row, these transformations have rich geometric content worth savoring before abstraction.

The simplest such transformations scale space uniformly. A matrix cI multiplies each coordinate by the scalar c, dilating or contracting space about the origin. More interesting are matrices that scale different directions differently:

$$\begin{bmatrix} 2 & 0 \\ 0 & 1/2 \end{bmatrix} \begin{pmatrix} x \\ y \end{pmatrix} = \begin{pmatrix} 2x \\ y/2 \end{pmatrix}$$

Such transformations stretch space along one axis while compressing along another – like a funhouse mirror's distortion rendered precise in coordinates.

Rotations in the plane arise from matrices of the form

$$\begin{bmatrix} \cos\theta & -\sin\theta \\ \sin\theta & \cos\theta \end{bmatrix}$$

spinning vectors through angle θ counterclockwise about the origin. That such matrices preserve lengths and angles is a consequence of their structure – see Chapter 5.

More subtle are *shear transformations*, such as

$$\begin{bmatrix} 1 & h \\ 0 & 1 \end{bmatrix}$$

which offset each horizontal line by an amount proportional to its height. These preserve area while tilting vertical lines – like a deck of cards carefully slid across a table.

These elementary transformations – scaling, rotation, and shear – combine to generate all linear transformations in the plane. Any 2×2 matrix can be understood as a composition of such basic geometric operations. This decomposition previews deeper structure to come, when we learn to factor matrices into simpler constituent parts.

What features do these transformations share, beyond their realization through matrix multiplication? First, they preserve the origin – the zero vector remains fixed. Second, they respect vector addition: the image of a sum equals the sum of the images. Third, they interact naturally

Question: What happens with the transpose of this horizontal shear?

with scalar multiplication: doubling an input vector doubles its image. These properties – seemingly obvious in the matrix setting – will form the scaffolding for our abstract theory.

Consider as well what these transformations can destroy. A rotation preserves distances but changes coordinates. A shear preserves areas but distorts angles. A scaling changes both distances and areas, but preserves lines through the origin. This selective preservation of geometric features hints at deeper invariants – quantities or properties that remain unchanged under certain classes of transformations.

The matrix transformation Ax converts geometric intuition about transforming space into algebraic manipulation of coordinates. As we lift these ideas to abstract vector spaces, this interplay between geometry and algebra will remain. Though we may lose the ability to visualize transformations directly, the core ideas – preservation of vector operations, study of invariants, decomposition into simpler parts – will guide our development.

Foreshadowing: The properties we observe in matrix transformations – preservation of vector operations – will define linearity in the abstract setting.

3.2   *Definitions & Implications*

Our experience with Euclidean transformations suggests key features that characterize the essence of linearity: preservation of addition and scaling. Many important transformations share these algebraic properties while lacking obvious geometric interpretation. This motivates abstracting away from geometry to study linear transformations between arbitrary vector spaces.

Definition 3.1 (Linear Transformation). Let V and W be vector spaces. A *linear transformation $T : V \to W$* is a function satisfying two properties:

1. Additivity: $T(v_1 + v_2) = T(v_1) + T(v_2)$ for all $v_1, v_2 \in V$
2. Homogeneity: $T(cv) = cT(v)$ for all $c \in \mathbb{R}$ and $v \in V$

These two properties combine to ensure that linear transformations preserve linear combinations. This seemingly simple requirement has profound implications.

Lemma 3.2. *A linear transformation $T : V \to W$ satisfies:*

1. $T(0) = 0$
2. $T(-v) = -T(v)$ *for all* $v \in V$
3. *T preserves linear combinations*

$$T\left(\sum_{i=1}^{n} c_i v_i\right) = \sum_{i=1}^{n} c_i T(v_i)$$

The preservation of linear combinations has an important consequence for subspaces: linear transformations send subspaces to subspaces.

Lemma 3.3. *If $T : V \to W$ is linear and $U < V$, then $T(U) < W$.*

Foreshadowing: The preservation of linear combinations will be the key to understanding how linear transformations interact with bases and coordinate systems.

Examples of Linear Transformations

The simplest examples of linear transformations are those of the previous section – any matrix A acts on Euclidean vectors via the linear transformation $T_A(x) = Ax$. This is so matural that it is hardly worth calling out as anything different than the matrix itself. However, not all linear transformations are so explicit in coordinates. Consider the following examples of a less geometric nature.

Example 3.4 (Differentiation). Consider the differentiation operator $D = d/dx$ from calculus. This satisfies linearity in that $D(f + g) = Df + Dg$ for differentiable functions f and g; and $D(cf) = c\,Df$ for c a scalar. As such, it defines a linear transformation from $C^\infty(\mathbb{R})$ to itself. If finite-dimensional vector spaces are preferred, one can restrict to polynomials, in which case $D : \mathcal{P}_n \to \mathcal{P}_{n-1}$:

Note that the subspace of constant polynomials is sent to zero. What happens to other subspaces of $\mathcal{P}_n$?

$$D\left(\sum_{i=0}^{n} c_i x^i\right) = \left(\sum_{j=1}^{n} j c_j x^{j-1}\right).$$

Here we see linearity without geometry – derivatives of sums are sums of derivatives, and constants factor out of derivatives.   ◇

Example 3.5 (Integration). The definite integration operator $I : C([a, b]) \to \mathbb{R}$ defined by

$$I(f) = \int_a^b f(x)\,dx$$

is, like differentiation, linear. What happens when we restrict attention to polynomials and to subspaces of polynomials?   ◇

The relationship between linear transformations and linear independence cuts to the heart of their structure.

Definition 3.6 (Injective and Surjective). A linear transformation $T : V \to W$ is:

1. *injective* (or *one-to-one*) if distinct inputs yield distinct outputs: $T(v_1) = T(v_2)$ implies $v_1 = v_2$
2. *surjective* (or *onto*) if every vector in W is the image of some vector in V: for each $w \in W$ there exists $v \in V$ such that $T(v) = w$

Lemma 3.7. *For a linear transformation* $T : V \to W$:

1. T *is injective if and only if it preserves linear independence*
2. T *is surjective if and only if* $T(V) = W$
3. T *is invertible if and only if it is both injective and surjective*

The power of these abstractions lies in their breadth of application. Whether transforming geometric vectors, polynomials, or functions, the same principles govern their behavior.

Caveat: A linear transformation can fail to be invertible in two distinct ways: by mapping different vectors to the same image (non-injective) or by missing vectors in the target space (non-surjective).

3.3 Isomorphisms

When are two vector spaces fundamentally the same? The geometric vectors in $\mathbb{R}^2$ seem quite different from the linear polynomials $ax + b$, yet both allow the same operations and satisfy the same rules. Such observations lead us to examine what it means for vector spaces to be indistinguishable from the perspective of linear algebra.

Linear transformations are directed mappings between to vector spaces. Given $T : V \to W$, we call V the *domain* and W the *codomain*. To serve as a perfect dictionary between spaces, T must possess both properties introduced in Definition 3.6: it must be both injective and surjective. Such transformations are called isomorphisms:

Nota bene: The use of *codomain* may be unfamiliar, as it stems from category theory. Such is also the case with the terms *monomorphism* and *epimorphism* for injective and surjective maps respectively, though we shall not use those particular terms.

Definition 3.8 (Isomorphism). Vector spaces V and W are *isomorphic*, denoted $V \cong W$, if there exists an *isomorphism* between them – a linear transformation that is both injective and surjective.

Example 3.9 (Coordinate vectors). The transformation $T : \mathbb{R}^2 \to \mathcal{P}_1$ sending vectors to linear polynomials via

$$\begin{pmatrix} a \\ b \end{pmatrix} \mapsto a + bx$$

is an isomorphism. It provides a perfect dictionary between geometric vectors and linear polynomials, preserving all vector space operations. Addition of vectors corresponds to addition of polynomials; scaling vectors means scaling polynomials. ◇

Example 3.10 (Matrix representations). The space $\mathbb{R}^{2\times 2}$ of 2-by-2 matrices is isomorphic to $\mathbb{R}^4$ via the transformation

$$\begin{bmatrix} a & b \\ c & d \end{bmatrix} \mapsto \begin{pmatrix} a \\ b \\ c \\ d \end{pmatrix}$$

Though we typically write matrices in square array, they are fundamentally no different from vectors – they simply package the same information differently.

Nota bene: Matrix multiplication is invisible to the vector space structure.

Not all linear transformations achieve this perfect correspondence. The projection $\Pi : \mathbb{R}^3 \to \mathbb{R}^2$ given by

$$\Pi \begin{pmatrix} x \\ y \\ z \end{pmatrix} = \begin{pmatrix} x \\ y \end{pmatrix}$$

is surjective but not injective – it reaches every point in the plane but collapses all points differing only in their z-coordinate. Conversely, the embedding $\iota : \mathbb{R}^2 \to \mathbb{R}^3$ given by

$$\iota \begin{pmatrix} x \\ y \end{pmatrix} = \begin{pmatrix} x \\ y \\ 0 \end{pmatrix}$$

is injective but not surjective – it preserves all information about vectors in the plane but misses most of $\mathbb{R}^3$.

Question: Are any two vector spaces with the same dimension in fact isomorphic?

These examples suggest a deep truth: vector spaces of different dimensions cannot be isomorphic. The projection above shows that a larger space cannot inject into a smaller one without collapsing; the embedding shows a smaller space cannot surject onto a larger one without missing vectors. This observation – though as yet unproven – hints at the fundamental nature of dimension in linear algebra.

The language of isomorphisms provides more than mere classification – it offers a perspective on what features of vector spaces truly matter. When spaces are isomorphic, we may freely translate problems between them, choosing whichever representation is most convenient. The geometric intuition of $\mathbb{R}^n$ becomes available to spaces of polynomials, matrices, or signals, provided we have constructed the right dictionary between them.

Example 3.11 (Polynomial derivatives). The differentiation operator $D : \mathcal{P}_2 \to \mathcal{P}_1$ given by

$$D(ax^2 + bx + c) = 2ax + b$$

is surjective but not injective. Every linear polynomial is a derivative (surjective), but constants vanish under differentiation (non-injective).

3.4 *Image & Kernel*

The subspaces we encountered in Chapter 2 arose from operations within a single vector space. Linear transformations generate their own characteristic subspaces – in both domain and codomain. These subspaces capture the essential features of how the transformation acts, measuring both its effectiveness and its defects.

Fix throughout a linear transformation $T : V \to W$ between vector spaces. The first subspace of interest lies in the codomain.

Definition 3.12 (Image). The *image* of T, denoted $\operatorname{im} T$, is the subspace of the codomain consisting of all possible outputs:

$$\operatorname{im} T = \{T(v) : v \in V\} < W \tag{3.1}$$

That this is indeed a subspace of W follows readily: the zero vector is certainly in the image (as $T(0) = 0$), and if $T(v_1)$ and $T(v_2)$ are any vectors in the image, then their sum $T(v_1) + T(v_2) = T(v_1 + v_2)$ is also in the image, as is any scalar multiple. The image measures the "reach" of the transformation – how much of W can be attained as output.

Definition 3.13 (Kernel). The *kernel* (or *nullspace*) of T, denoted $\ker T$, is the subspace of the domain consisting of all vectors that vanish under T:

$$\ker T = \{v \in V : T(v) = 0\} < V \tag{3.2}$$

That this too forms a subspace of V is again straightforward: the zero vector is certainly in the kernel; and if v_1 and v_2 are in the kernel, then $T(v_1 + v_2) = T(v_1) + T(v_2) = 0$, with scalar multiples following similarly. The kernel captures what the transformation cannot "see" – the vectors that disappear under its action.

These abstract definitions crystallize our earlier work with linear systems. Consider the matrix equation $Ax = b$ from Chapter 1. This defines a linear transformation $T_A : \mathbb{R}^n \to \mathbb{R}^m$ via $T_A(x) = Ax$. The standard questions about this system now have geometric meaning:

1. Does a solution exist? Yes if and only if $b \in \operatorname{im} T_A$.
2. Is the solution unique? Yes if and only if $\ker T_A = \{0\}$.
3. If more than one solution exists, how are they related? They differ by elements of $\ker T_A$.

Example 3.14 (Calculus operators). The differentiation operator $D : C^1([a,b]) \to C([a,b])$ has kernel consisting of all constant functions on

$[a, b]$ – these are precisely the functions that vanish under differentiation. Its image consists of all continuous functions that arise as derivatives, a proper subspace of $C([a, b])$ (not every continuous function is a derivative).

The definite integration operator $I : C([a, b]) \to \mathbb{R}$ defined by $I(f) = \int_a^b f(x)dx$ has (very large) kernel consisting of all functions whose integral over $[a, b]$ vanishes. Its image is all of $\mathbb{R}$ – any real number can be realized as the integral of some continuous function. ◇

These subspaces provide the first tools for analyzing the structure of linear transformations. A transformation is one-to-one precisely when its kernel contains only the zero vector; it is onto when its image is the entire codomain. The interplay between these subspaces – how their dimensions balance, how they decompose the spaces involved – leads to the deeper theory ahead.

Foreshadowing: The relationship between the dimensions of kernel and image will prove fundamental to understanding linear transformations. This balance between what vanishes and what is attained is captured in the Fundamental Theorem at the end of this chapter.

3.5   Rank & Nullity

The dimension of a vector space captures its size and complexity. For a linear transformation, we seek similar measures of size and complexity – not of a single space, but of how the transformation acts between spaces. These measures arise naturally from the dimensions of image and kernel.

Definition 3.15 (Rank & Nullity). The *rank* of a linear transformation $T : V \to W$ is the dimension of its image:

$$\operatorname{rank} T = \dim(\operatorname{im} T) \tag{3.3}$$

The *nullity* of T is the dimension of its kernel:

$$\operatorname{null} T = \dim(\ker T) \tag{3.4}$$

This abstracts the pedestrian notion of matrix rank from Definition 1.10. When T is represented by a matrix A, the abstract and concrete ranks coincide.

Think: rank measures the transformation's effective dimension – like light passing through crystal, part transmitted and part absorbed, it counts the independent directions that survive. The nullity complements this measure by counting dimensions lost, those directions that, like perfectly absorbed light, vanish entirely in transformation.

Example 3.16 (Matrix Rank and Nullity). For a 3×4 matrix A of rank 2, the transformation $T_A : \mathbb{R}^4 \to \mathbb{R}^3$ has:

1. $\operatorname{rank} T_A = 2$, meaning $\operatorname{im} T_A$ is a plane in $\mathbb{R}^3$
2. $\operatorname{null} T_A = 2$, as solving $Ax = 0$ yields a two-dimensional solution space
3. $\dim(\ker T_A) + \dim(\operatorname{im} T_A) = \dim(\mathbb{R}^4) = 4$

This last observation hints at a deeper relationship between rank and nullity. ◇

Example 3.17 (Calculus operators). The differentiation operator $D : \mathcal{P}_n \to \mathcal{P}_{n-1}$ has:

1. $\operatorname{rank} D = n$, as every polynomial in $\mathcal{P}_{n-1}$ is a derivative
2. $\operatorname{null} D = 1$, as only constant functions vanish under differentiation
3. $\dim(\ker D) + \dim(\operatorname{im} D) = \dim(\mathcal{P}_n) = n + 1$

Again we see the dimensions balance. $\diamond$

Example 3.18 (Integration). Consider the definite integration operator $I : C[0,1] \to \mathbb{R}$ defined by $I(f) = \int_0^1 f(x)dx$. Though $C[0,1]$ is infinite-dimensional:

1. $\operatorname{rank} I = 1$, as the image is all of $\mathbb{R}$
2. $\operatorname{null} I$ is infinite-dimensional, containing all functions whose integral vanishes
3. The dimensional balance breaks down in the infinite-dimensional setting

$\diamond$

These examples suggest deep connections between rank, nullity, and the dimensions of domain and codomain.

Caveat: The relationship between rank and nullity becomes more subtle in infinite dimensions. The examples here are meant to build intuition in the finite-dimensional case.

3.6 Quotients

Linear transformations reveal structure not only through what they preserve, but through what they collapse. The manner in which different vectors map to identical outputs suggests a natural organization – grouping vectors that transform identically. This insight leads to one of the most profound constructions in linear algebra: the quotient space.

Definition 3.19 (Quotient Space). Let V be a vector space and $U < V$ a subspace. The *quotient space* V/U is the vector space whose elements are equivalence classes of vectors in V, where vectors $v_1, v_2 \in V$ are equivalent if and only if their difference lies in U:

$$v_1 \sim v_2 \iff v_1 - v_2 \in U$$

The equivalence class of $v \in V$, denoted $[v]$, consists of all vectors equivalent to v:

$$[v] = \{w \in V : w - v \in U\} = v + U$$

Vector operations on V/U are defined through representatives: $[v_1] + [v_2] = [v_1 + v_2]$ and $c[v] = [cv]$ for scalar c. $\bullet$

The physical intuition for quotient spaces emerges naturally in electrical networks. Consider a circuit with n nodes, where we measure voltage

The properties of subspaces ensure $\sim$ defines a proper equivalence relation: reflexive ($v \sim v$), symmetric ($v_1 \sim v_2$ implies $v_2 \sim v_1$), and transitive ($v_1 \sim v_2$ and $v_2 \sim v_3$ implies $v_1 \sim v_3$).

differences between pairs of nodes. Though each node has its own voltage potential, the physically meaningful measurements are always differences – adding a constant voltage to every node leaves all measurements unchanged. This observation reveals the fundamental role of quotients in physics.

Let $V = \mathbb{R}^n$ be the vector space of voltage assignments to nodes. The projection Π that sends each voltage configuration to its equivalence class under constant shifts:

$$\Pi : V \to Q \quad : \quad v \mapsto [v]$$

has kernel consisting precisely of constant voltage shifts $(c, c, \ldots, c)^T$. The quotient space $V / \ker(\Pi)$ then captures the physically meaningful voltage states, stripped of their artificial dependence on reference potential.

Example 3.20 (Kernel Quotients). Let $T : V \to W$ be a linear transformation. Two vectors that differ by an element of $\ker T$ are sent to the same output:

$$T(v_1) = T(v_2) \iff v_1 - v_2 \in \ker T$$

The quotient space $V / \ker T$ naturally represents the "effective" input space of T – it identifies inputs that T cannot distinguish. $\diamond$

The quotient space inherits a vector space structure from V. Addition and scalar multiplication are defined on equivalence classes:

$$[v_1] + [v_2] = [v_1 + v_2] \quad \text{and} \quad c[v] = [cv]$$

These operations are well-defined – independent of which representatives we choose from the equivalence classes.

Example 3.21 (Geometric quotients). Consider first quotienting $\mathbb{R}^3$ by a line L through the origin. Two points $p, q \in \mathbb{R}^3$ belong to the same equivalence class precisely when their difference $p - q$ lies in L – that is, when they differ by some vector parallel to L. Each equivalence class thus forms a plane parallel to L, as shifting along L keeps us within the same class. The quotient space $\mathbb{R}^3 / L$ can be visualized as the collection of all such parallel planes, naturally parametrized by a two-dimensional plane perpendicular to L. Though abstract, this quotient has dimension exactly 2, as specifying a plane requires two coordinates once we fix its direction.

Now consider instead quotienting $\mathbb{R}^3$ by a plane P through the origin. The equivalence classes are lines perpendicular to P – each point equivalent to all others lying directly above or below it relative to P. The quotient space $\mathbb{R}^3 / P$ becomes naturally one-dimensional, parametrized by

Foreshadowing: When we introduce inner products, each equivalence class will have a unique representative orthogonal to the subspace being quotiented. For now, we work with the classes themselves.

the signed distance along the normal vector to P. This distance provides a concrete realization of the abstract quotient: two points are equivalent precisely when they have the same projection onto the normal vector to P, or equivalently, when they lie the same distance from P along parallel perpendicular lines. ◇

The dimension of a quotient space reflects both the dimension of the original space and the dimension of the subspace being quotiented:

$$\dim(V/U) = \dim V - \dim U$$

This dimensional relationship will prove crucial in understanding how linear transformations decompose spaces.

Example 3.22 (Integration quotients). Consider the indefinite integral operator (or *antidifferentiation*) D^{-1} acting on continuous functions $C([a,b])$ on an interval. An antiderivative always exists (thanks to the FTIC) but is well-defined only up to a constant.

You did not forget the $+C$ did you?

To make this a linear transformation of vector spaces requires using a quotient space for the codomain. Let $U < C([a.b])$ denote the subspace of constant functions on the interval. Then the quotient $C([a,b])/U$ consists of classes of functions whose differences are constants. Indefinite integration is now a linear transformation

$$D^{-1} : C([a,b]) \to C([a,b])/U.$$

◇

Example 3.23 (Translation Invariance). Consider a dataset of n points in $\mathbb{R}^d$, represented as columns of a matrix $X \in \mathbb{R}^{d\times n}$. In many applications, such as clustering or pattern recognition, we care about the relative positions of points rather than their absolute positions in space.

Let $V = \mathbb{R}^{d\times n}$ be the vector space of all possible datasets. The subspace of uniform translations $U < V$ consists of all matrices whose columns are identical:

$$U = \left\{ \begin{bmatrix} v & v & \cdots & v \end{bmatrix} : v \in \mathbb{R}^d \right\} = \{ v \cdot \mathbf{1}^T : v \in \mathbb{R}^d \}$$

where $\mathbf{1} \in \mathbb{R}^n$ is the vector of all ones. Each matrix in U represents a uniform translation of the zero matrix by some vector $v \in \mathbb{R}^d$.

Two datasets $X, Y \in V$ are equivalent modulo translation if their difference $X - Y$ lies in U – that is, if there exists some vector $v \in \mathbb{R}^d$ such that each column of X is translated by v to obtain the corresponding column of Y:

$$X \sim Y \iff X - Y \in U \iff X - Y = v \cdot \mathbf{1}^T \text{ for some } v \in \mathbb{R}^d$$

The quotient space V/U then represents datasets modulo translation –
it captures the intrinsic shape of point configurations while ignoring their
absolute position in space. A dimension count reveals the structure:

- $\dim(V) = dn$ (coordinates of n points in $\mathbb{R}^d$)
- $\dim(U) = d$ (coordinates of the translation vector)
- $\dim(V/U) = dn - d$ (data sets up to translation)

This gives an explicit isomorphism between V/U and the subspace of mean-zero datasets.

$\diamond$

Quotient spaces provide a formal way to identify vectors that behave
similarly under certain operations. When we quotient a domain by the
kernel of a transformation, we obtain a space that faithfully represents
how the transformation acts, stripped of redundancy. This perspective
will prove invaluable as we develop more sophisticated tools for analyzing linear transformations.

3.7   Coimage & Cokernel

The image and kernel of a linear transformation tell only half the story.
Just as quotient spaces reveal structure by identifying vectors that behave similarly, we can illuminate the action of a linear transformation by
examining quotients in both domain and codomain. This leads to two
additional spaces that complete our structural understanding.

Definition 3.24 (Coimage). Given a linear transformation $T : V \to W$, the
coimage of T is the quotient space

$$\operatorname{coim} T = V/\ker T \tag{3.5}$$

The coimage represents the "effective" input space of T – it identifies
inputs that T cannot distinguish. The projection $\Pi : V \to \operatorname{coim} T$ sends
each vector to its equivalence class modulo $\ker T$.

The coimage has a natural interpretation: it measures how many independent input directions actually influence the output. Two vectors in
V produce the same output precisely when they differ by an element of
$\ker T$ – thus $\operatorname{coim} T$ parametrizes the truly distinct inputs as seen by T.

Example: For a rank-2 matrix $A : \mathbb{R}^4 \to \mathbb{R}^3$, the coimage is 2-dimensional, representing the two independent input directions that affect the output.

Our fourth fundamental space is the most hidden and obscure:

Definition 3.25 (Cokernel). Given a linear transformation $T : V \to W$, the
cokernel of T is the quotient space

$$\operatorname{coker} T = W/\operatorname{im} T \tag{3.6}$$

The cokernel measures the failure of T to reach all of W. When T is surjective, coker T is trivial; otherwise, it captures the "invisible" space in the codomain.

Example 3.26 (Differential operators). For the derivative operator $D : \mathcal{P}_n \to \mathcal{P}_{n-1}$:

1. The coimage is (n)-dimensional, as only constants vanish under D
2. The cokernel is trivial, as every polynomial in $\mathcal{P}_{n-1}$ is a derivative

For integration $I : C[0,1] \to \mathbb{R}$:

1. The coimage captures functions differing by more than their average
2. The cokernel is trivial, as every real number is an integral

$\diamond$

> *Foreshadowing:* The dimensions of these spaces are not independent – they satisfy a beautiful relationship that the next section will reveal.

These four fundamental spaces – kernel, image, cokernel, and coimage – form a complete set of structural invariants for a linear transformation. Their relationships, seemingly complex, crystallize in the Fundamental Theorem ahead. Each measures a different aspect of how T transforms space:

- ker T captures what vanishes
- im T shows what is attainable
- coim T reveals independent inputs
- coker T measures missing outputs

The stage is now set for a deeper understanding of how these spaces fit together – a unification that explains not just what these spaces are, but why they must exist and how they relate.

3.8   The Fundamental Theorem

Mathematics achieves clarity through unification. The various structures we have encountered – images and kernels, quotients, cokernels and coimages – are not merely related but fundamentally interwoven. The tapestry they form is one of the most beautiful results in linear algebra, revealing deep connections between algebraic and dimensional properties of linear transformations.

Consider a linear transformation $T : V \to W$ between finite-dimensional vector spaces. From our explorations, we have uncovered four fundamental subspaces:

1. The image im T, capturing all possible output
2. The kernel ker T, containing all inputs that vanish
3. The coimage coim T, encoding independent input directions
4. The cokernel coker T, measuring the failure to surject

These spaces, seemingly distinct, are bound together by a profound theorem that explains not only how they relate but why they must so relate. This is the Fundamental Theorem of Linear Algebra:

Theorem 3.27 (Fundamental Theorem of Linear Algebra). *For any linear transformation $T : V \to W$ between finite-dimensional vector spaces, the following relationships hold and are equivalent:*

1. *The domain and codomain decompose as direct sums:*

$$V \cong \ker T \oplus coim\, T \quad and \quad W \cong im\, T \oplus coker\, T$$

2. *The coimage and image are naturally isomorphic:* $coim\, T \cong im\, T$

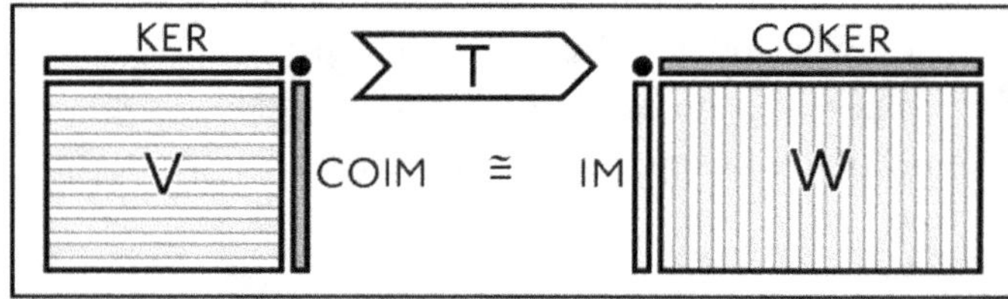

When translated to dimensions, the Fundamental Theorem is sometimes called the *Rank-Nullity Theorem*:

Corollary 3.28 (Rank-Nullity). *For a linear transformation between finite-dimensional vector spaces, the dimensions balance in complementary pairs:*

$$\dim V = \dim(\ker T) + \dim(coim\, T) \quad and \quad \dim W = \dim(im\, T) + \dim(coker\, T)$$

Furthermore, the rank connects domain and codomain:

$$\dim(coim\, T) = rank(T) = \dim(im\, T)$$

Otherwise said:

$$\dim V = null(T) + rank(T)$$

Example 3.29 (Matrix rank). When T is represented by a matrix A, these relationships explain why:

1. The nullity $null(A)$ plus the rank $rank(A)$ equals the number of columns of A
2. The *row rank* $(= \dim \textsc{row}(A))$ and the *column rank* $(= \dim \textsc{col}(A))$ are equal to $rank(A)$

These familiar facts from matrix algebra are manifestations of the deeper structural relationships guaranteed by the Fundamental Theorem.   ◇

Example: For the projection $T : \mathbb{R}^3 \to \mathbb{R}^2$ onto the xy-plane, the kernel is the z-axis, the image is $\mathbb{R}^2$, the coimage is the xy-plane in the domain, and the cokernel is trivial.

"Four mighty ones there are in every Man; a Perfect Unity"

Foreshadowing: These algebraic relations will acquire additional geometric significance when we introduce inner products in Chapter 5.

The theorem has immediate practical implications. When solving a linear system $Tx = b$, we now understand that:

1. A solution exists if and only if $b \in \text{im}(T)$
2. When a solution exists, others differ by elements of $\ker(T)$
3. Any solution can be uniquely decomposed into parts from $\text{coim}(T)$ and $\ker(T)$
4. The obstruction to existence lies in $\text{coker}(T)$

Example 3.30 (Solving linear systems). Consider solving $Ax = b$ where A is 3×4 of rank 2. The Fundamental Theorem tells us that:

1. The nullspace has dimension 2
2. The image has dimension 2
3. Solutions exist only when b lies in a 2-dimensional subspace
4. When solutions exist, they form a 2-dimensional affine space

$\diamond$

Foreshadowing: When we introduce inner products, these decompositions will provide the foundation for finding optimal approximate solutions when exact solutions do not exist.

The Fundamental Theorem is more than a collection of relationships – it is a lens through which we view linear transformations. Whether analyzing electrical networks, processing signals, or fitting models to data, these structural relationships guide our understanding and inform our computations. The decompositions it guarantees will prove even more powerful when enhanced with geometric structure in subsequent chapters.

Graph Topology & Network Structure

When a city's power grid fails, engineers must quickly identify which neighborhoods remain connected and where backup pathways exist. Similar questions arise across networks: Can a signal reach all neurons in a circuit? Will information flow reliably through a social network? Does a computer network contain redundant paths to route around failures? These practical concerns about connectivity and resilience share a deep mathematical structure that emerges through careful application of linear algebra to network topology.

Consider a finite directed graph $G = (V, E)$ with vertex set $V = \{v_1, \ldots, v_n\}$ and edge set $E = \{e_1, \ldots, e_m\}$. Each edge e has a specified orientation, with starting vertex e^- and ending vertex e^+. From this discrete structure we construct two fundamental vector spaces:

- $C_0(G)$: the vector space with basis elements the vertices V
- $C_1(G)$: the vector space with basis elements the oriented edges E

These spaces have dimensions n and m respectively. Though their elements can be interpreted as assignments of numbers to vertices or edges (like voltages or currents), viewing them as abstract vector spaces clarifies their fundamental structure.

The relationship between these spaces emerges through a natural transformation called the *boundary operator* $\partial : C_1(G) \to C_0(G)$. On basis elements, this operator acts by:

$$\partial(e) = e^+ - e^-$$

extending linearly to all of $C_1(G)$. Though defined using edge orientations, its fundamental properties — captured through kernel and cokernel — prove independent of these choices.

Example 3.31 (Ladder Network). Consider a "ladder" network with eight vertices and ten edges arranged and labeled as in the figure, right. The boundary operator has matrix representation:

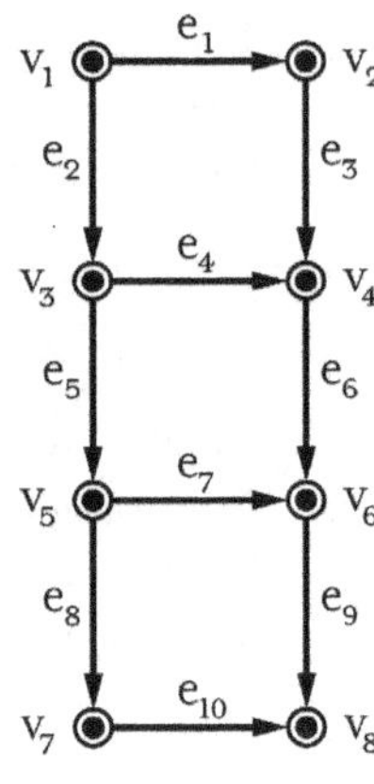

$$\partial = \begin{bmatrix}
-1 & -1 & 0 & 0 & 0 & 0 & 0 & 0 & 0 & 0 \\
1 & 0 & -1 & 0 & 0 & 0 & 0 & 0 & 0 & 0 \\
0 & 1 & 0 & -1 & -1 & 0 & 0 & 0 & 0 & 0 \\
0 & 0 & 1 & 1 & 0 & -1 & 0 & 0 & 0 & 0 \\
0 & 0 & 0 & 0 & 1 & 0 & -1 & -1 & 0 & 0 \\
0 & 0 & 0 & 0 & 0 & 1 & 1 & 0 & -1 & 0 \\
0 & 0 & 0 & 0 & 0 & 0 & 0 & 1 & 0 & -1 \\
0 & 0 & 0 & 0 & 0 & 0 & 0 & 0 & 1 & 1
\end{bmatrix}$$

This network contains many cycles, yet there are but three that can be chosen to be *independent*. One simple representative follows the top square $e_2 + e_4 - e_3 - e_1$, where the minus signs indicate traversing an edge opposite its orientation. This, together with the other two obvious squares, forms a basis for $\ker \partial$. ◇

Such *cycles* — elements of $\ker(\partial)$ — represent closed paths through the network where the "flow" in equals flow out at each vertex. Not every element of $\ker(\partial)$ corresponds to a simple cycle; some represent combinations of cycles. The dimension of this kernel, denoted β_1, counts the number of *independent* cycles — those that cannot be expressed as combinations of smaller cycles. A power grid with larger β_1 offers more backup paths; a neural circuit with independent cycles can sustain more complex recurrent patterns.

The cokernel of ∂ reveals complementary structure through its quotient space $C_0(G)/\operatorname{im}(\partial)$. This space effectively identifies vertices that can be reached from each other through network paths. Its dimension β_0 counts the network's connected components. A power grid with $\beta_0 > 1$ has disconnected regions requiring immediate attention; a neural network with multiple components represents independent processing modules.

These numbers satisfy a remarkable relationship:

$$\beta_1 - \beta_0 = m - n \tag{3.7}$$

That is, the number of independent cycles minus the number of connected components equals the excess between edge count and vertex count. This equation — simultaneously the rank-nullity theorem for ∂ and a combinatorial invariant of the graph — expresses a fundamental balance between cycles and components. Adding edges tends to create cycles (β_1 increases) while joining components (β_0 decreases).

Nota bene: the symbol β stands for *Betti number,* a fundamental object of study in algebraic topology.

Question: what happens if you subdivide each edge, adding a new vertex in the midst of each edge, and splitting the edge in two?

Example 3.32 (Point Cloud Data). Modern data analysis often begins with points sampled from some underlying shape or manifold. Given points $\{x_1, \ldots, x_N\}$ in $\mathbb{R}^d$, we can construct a graph by connecting points within distance ε of each other. The resulting network's topology — measured through β_0 and β_1 — reveals fundamental features of the underlying data:

- β_0 counts clusters in the data
- β_1 detects holes and cycles
- The dependence of these numbers on ε characterizes scale

This forms the foundation of modern *Topological Data Analysis*, where persistence of these features across scales indicates genuine structure rather than noise. ◇

The practical value of this framework lies in its reduction of qualitative questions about network structure to systematic linear algebra. What began as abstract study of kernels and images emerges as practical tools for understanding networks. The fundamental theorem, revealing deep relationships between these spaces, provides the mathematical foundation for analyzing complex networks.

Exercises: Chapter 3

1. Let $T : \mathbb{R}^2 \to \mathbb{R}^2$ be the linear transformation defined by $T(x, y) = (2x + y, x - y)$. Find the kernel and image of T. Is T injective? Is T surjective? Find the rank and nullity of T.

2. Consider the differentiation operator $D : \mathcal{P}_2 \to \mathcal{P}_1$ defined by $D(ax^2 + bx + c) = 2ax + b$. Compute dimensions for $\ker D$ and $\operatorname{im} D$. Verify the rank-nullity theorem for this transformation.

3. For a fixed vector $v \in \mathbb{R}^n$, define $T_v : \mathbb{R}^n \to \mathbb{R}$ by $T_v(x) = v \cdot x$. Prove that this defines a linear transformation and find its kernel and image.

4. Let V be the vector space of 2×2 matrices. Define $T : V \to V$ by $T(A) = A^T$. Prove that T is linear. Find the kernel of T. Show that $T \circ T = I$ (where I is the identity transformation). Is T an isomorphism? Justify your answer.

5. Consider the mass operator $M : C([0, L]) \to \mathbb{R}$ defined by $M(f) = \int_0^L f(x)\, dx$, where f is interpreted as a linear density. Prove that M is a linear transformation. Is M surjective? Is the operation which computes center-of-mass a linear transformation?

6. Let $T : \mathbb{R}^3 \to \mathbb{R}^2$ be a linear transformation with $\ker T$ spanned by $(1, 0, -1)^T$. What is the dimension of $\operatorname{im} T$? Can T be onto? Find the dimension of $\operatorname{coim} T$.

7. Let $V = \mathbb{R}^2$ and define vectors $u, v \in V$ to be equivalent if they differ by a multiple of $a = (1, 1)^T$. Prove this is an equivalence relation. Describe geometrically what one equivalence class looks like. Show this equivalence relation corresponds to the quotient space V/U where $U = \operatorname{span}(a)$.

8. In $\mathbb{R}^3$, we say two vectors are equivalent if their first coordinates are equal. Show this defines an equivalence relation. Find a concrete description of the quotient space (hint: what is its dimension?). Describe the subspace that generates this equivalence relation.

9. For linear transformations $S, T : V \to W$, prove that $\operatorname{rank}(S + T) \leq \operatorname{rank}(S) +$

rank(T). When does equality hold?

10. Let $T : V \to W$ be linear. Prove that if $\{v_1, \ldots, v_k\}$ is linearly independent in V and $T(v_i) \neq \mathbf{0}$ for all i, then $\{T(v_1), \ldots, T(v_k)\}$ is linearly independent in W.

11. Let $T : V \to W$ be a linear transformation. Prove that any subspace of V containing $\ker T$ is mapped to a subspace of W of dimension at most $\operatorname{rank} T$.

12. Let V be finite-dimensional and $T : V \to V$ linear. Prove that T is invertible if and only if $T(\ker T) = \{\mathbf{0}\}$.

13. For linear transformations $S, T : V \to W$, prove that $\ker(S + T)$ contains $\ker S \cap \ker T$. Give an example where this containment is strict.

14. Let $C([0,1])$ be the vector space of continuous functions on $[0,1]$. We say two functions $f, g \in C([0,1])$ are equivalent if $f(0) = g(0)$. Prove this is an equivalence relation. Find the subspace that generates this equivalence relation. What is a simple way to describe the quotient space geometrically?

15. Let $v \in \mathbb{R}^n$ be nonzero. Two vectors $x, y \in \mathbb{R}^n$ are called equivalent if $v \cdot x = v \cdot y$. Prove this is an equivalence relation. Show that the quotient space defined by this equivalence relation is isomorphic to $\mathbb{R}$.

16. Let V be a vector space and $U < V$ a subspace. For $v \in V$, prove that the equivalence class $[v]$ in V/U equals the set $v + U = \{v + u : u \in U\}$. Use this to explain why vector addition in V/U is well-defined.

17. Let $T : V \to W$ be a linear transformation. Prove that the quotient space $V/\ker T$ is isomorphic to $\operatorname{im} T$ by explicitly constructing an isomorphism and verifying it is well-defined.

18. For a linear transformation $T : V \to W$, prove that $\dim(\ker T + \operatorname{im} T) = \dim V$ if and only if $T \circ T = T$.

19. If $S, T : V \to W$ are linear transformations and $\ker S = \ker T$, then there exists an isomorphism $\varphi : \operatorname{im} S \to \operatorname{im} T$.

20. Let V be a finite-dimensional vector space and $U < V$ a subspace. Prove that the quotient map $\pi : V \to V/U$ defined by $\pi(v) = [v]$ is a linear transformation. What is its kernel?

21. Let V be the vector space of 2×2 matrices. Define two matrices $A, B \in V$ to be equivalent if their traces are equal. Show this is an equivalence relation and identify the quotient space with a familiar vector space.

22. Consider the differentiation operator $D : \mathcal{P}_2 \to \mathcal{P}_1$ defined by $D(ax^2 + bx + c) = 2ax + b$. Find $\ker D$ and $\operatorname{im} D$. Verify the rank-nullity theorem for this transformation. Find an explicit formula for a linear transformation $S : \mathcal{P}_1 \to \mathcal{P}_2$ such that $D \circ S = I_{\mathcal{P}_1}$.

Chapter 4
Bases & Coordinates

"fixing them firm on their base, the bellows began to blowe"

THE PASSAGE FROM ABSTRACT TO CONCRETE is as important as its inverse. Having dwelt in the realm of abstract vector spaces and transformations, we now seek to make these concepts precise and measurable through coordinates and computations. This chapter bridges the gap between the theoretical structures we have built and the practical tools needed to work with them.

The concepts are familiar from geometric intuition: we routinely describe points in space using coordinates relative to chosen axes. These coordinates transform an abstract point into a concrete list of numbers that we can manipulate. Yet this simple idea – that we can systematically assign numbers to abstract vectors – contains surprising depth. The choice of coordinate system, seemingly arbitrary, can dramatically affect how easily we solve problems or understand structures.

This tension between intrinsic properties and their coordinate representations lies at the heart of linear algebra. A vector space exists independent of any particular way we choose to measure it, yet we cannot compute without making such choices. A linear transformation acts geometrically, yet we encode it as a matrix only after choosing bases for its domain and codomain. These choices – of bases and the coordinates they induce – form the bridge between abstract understanding and concrete computation.

Our task is to build this bridge carefully, ensuring it carries both theoretical insight and practical utility across the gap. We begin with the notion of a basis – a set of vectors that both spans a space and does so efficiently. These bases provide coordinate systems, allowing us to translate abstract vectors into concrete lists of numbers. The interplay between different bases leads us to change of coordinates formulas, revealing how geometric objects appear from different perspectives.

This systematic encoding of abstract structures into numerical form enables the algorithmic heart of linear algebra. The price we pay for such computational power is a proliferation of coordinate-dependent calculations. Our challenge is to maintain sight of coordinate-independent truths while wielding coordinates as practical tools. This interplay between abstract understanding and concrete computation will guide our development throughout this text.

4.1   Bases & Spanning Sets

From Chapter 2, recall that a spanning set for a vector space may contain redundant vectors, while a linearly independent set may fail to reach all vectors in the space. The notion of a basis synthesizes these concepts, providing a set of vectors that spans efficiently – without redundancy and without gaps.

Definition 4.1 (Basis). A *basis* for a vector space V is a set $\mathcal{B}$ of vectors that spans V yet is linearly independent. •

The economy of this definition belies its power. A basis provides a minimal spanning set – minimal in the sense that removing any vector from the basis leaves a set that no longer spans V. Equivalently, it provides a maximal linearly independent set – maximal in that adding any vector creates linear dependence.

Example 4.2 (Polynomial bases). The space $\mathcal{P}_2$ of quadratic polynomials admits several natural choices of basis, each providing different advantages:

1. The monomial basis $\{1, x, x^2\}$
2. The Lagrange basis for interpolation points $\{-1, 0, 1\}$:

$$\left\{ \frac{x(x+1)}{2}, -x^2 + 1, \frac{x(x-1)}{2} \right\}$$

3. The Newton basis with nodes d and e $(d \neq e)$:

$$\{1, (x - d), (x - d)(x - e)\}$$

To illustrate, consider the polynomial $p(x) = x^2 + x + 1$. In the monomial basis, it's already in standard form:

$$p(x) = 1 \cdot 1 + 1 \cdot x + 1 \cdot x^2$$

In the Lagrange basis at points $\{-1, 0, 1\}$, writing out the expansion:

$$p(x) = 2 \cdot \frac{x(x+1)}{2} + 1 \cdot (-x^2 + 1) + 2 \cdot \frac{x(x-1)}{2} = x^2 + x + 1$$

This demonstrates how different bases can represent the same polynomial in ways that are advantageous for different computational purposes. ◇

The existence of a basis for any vector space is guaranteed by the following result, whose proof illuminates the relationship between spanning and independence:

Theorem 4.3 (Basis Extension). *Let V be a finite-dimensional vector space and $S \subseteq V$ be a linearly independent set. Then S can be extended to a basis of V by adding finitely many vectors. Moreover, if $\dim V = n$ and $|S| = k \leq n$, then exactly $n - k$ vectors need to be added.*

Proof. Let $\{v_1, \ldots, v_n\}$ be any basis of V. We can extend S by adding vectors from this basis one at a time until we obtain a basis, stopping when we achieve a spanning set. The process must terminate after at most $n - k$ steps since any linearly independent set in V has size at most n. □

This constructive proof reveals a fundamental principle: we can build bases either by extension (adding vectors until we span) or by reduction (removing vectors until independence). Both processes terminate because of finite-dimensionality – a crucial hypothesis that fails in infinite-dimensional spaces.

Example 4.4 (Matrix bases). The space $\mathbb{R}^{2\times 2}$ of 2×2 matrices has the standard basis

$$E_{11} = \begin{bmatrix} 1 & 0 \\ 0 & 0 \end{bmatrix}, \quad E_{12} = \begin{bmatrix} 0 & 1 \\ 0 & 0 \end{bmatrix}, \quad E_{21} = \begin{bmatrix} 0 & 0 \\ 1 & 0 \end{bmatrix}, \quad E_{22} = \begin{bmatrix} 0 & 0 \\ 0 & 1 \end{bmatrix}$$

This basis makes the coordinate structure transparent but obscures other properties. For example, the basis

$$\begin{bmatrix} 1 & 0 \\ 0 & 1 \end{bmatrix}, \quad \begin{bmatrix} 0 & 1 \\ 1 & 0 \end{bmatrix}, \quad \begin{bmatrix} 0 & -1 \\ 1 & 0 \end{bmatrix}, \quad \begin{bmatrix} 1 & 0 \\ 0 & -1 \end{bmatrix}$$

better reveals the decomposition into symmetric and skew-symmetric parts. ◇

A key property of bases is that they all have the same size, thanks to Lemma 2.22:

Foreshadowing: Different bases reveal different aspects of a space's structure. In Chapter 7, we shall discover bases that illuminate the action of linear differential equations.

The theorem generalizes to infinite-dimensional spaces using Zorn's Lemma, though the extension process becomes non-constructive.

Caveat: The process of extending to a basis or extracting one from a spanning set is not unique – different choices yield different bases.

Corollary 4.5. *Any two bases of a vector space have the same number of vectors.*

This reveals dimension as an intrinsic property of the space, independent of choice of basis.

Example 4.6 (Dimension counting). The following dimensions arise naturally:

1. $\dim(\mathbb{R}^n) = n$
2. $\dim(\mathcal{P}_n) = n + 1$
3. $\dim(\mathbb{R}^{m \times n}) = mn$
4. $\dim(\text{SYM}_n) = \frac{1}{2}n(n+1)$

Each counts the minimal number of parameters needed to specify an element of the space.

Recall, $\mathcal{P}_n$ is the space of polynomials of degree $\leq n$ and SYM_n denotes the space of symmetric n-by-n matrices.

$\diamond$

Bases provide our first systematic way to measure vector spaces. The choice of basis – which is always somewhat arbitrary – trades the intrinsic nature of the space for concrete computability. This tension between coordinate-free properties and coordinate-dependent calculations will be a recurring theme as we develop the machinery of linear algebra.

4.2   Coordinates & Components

The existence of a basis provides more than a spanning set for a vector space – it enables a systematic translation of abstract vectors into concrete lists of numbers. This translation, seemingly mundane, is the keystone of computational linear algebra. It bridges the gap between geometric intuition and algorithmic manipulation.

The crucial observation is that any vector in a space can be written uniquely as a linear combination of basis vectors. Given a basis $\{b_1, \ldots, b_n\}$ for a vector space V, each vector $v \in V$ has a unique expression:

$$v = c_1 b_1 + c_2 b_2 + \cdots + c_n b_n$$

The scalars $c_1, \ldots, c_n$ are called the *coordinates* of v relative to this basis. The ordered list of these coordinates, written as a column vector

Caveat: The notation $[v]_{\mathcal{B}}$ emphasizes that coordinates depend on choice of basis. A vector has different coordinates in different bases, though the vector itself remains unchanged.

$$[v]_{\mathcal{B}} = \begin{pmatrix} c_1 \\ c_2 \\ \vdots \\ c_n \end{pmatrix}$$

is the *coordinate vector* of v with respect to the basis $\mathcal{B} = \{b_1, \ldots, b_n\}$.

Example 4.7 (Polynomial coordinates). Consider the polynomial $p(x) = 6 + 2x - 3x^2$ in $\mathcal{P}_2$. In the monomial basis $\mathcal{M} = \{1, x, x^2\}$, its coordinate vector is

$$[p]_{\mathcal{M}} = \begin{pmatrix} 6 \\ 2 \\ -3 \end{pmatrix}$$

In the Lagrange basis $\mathcal{L} = \{1, x - 1, (x - 1)(x + 1)\}$, the same polynomial has different coordinates:

$$[p]_{\mathcal{L}} = \begin{pmatrix} 6 \\ 2 \\ -3/2 \end{pmatrix}$$

The polynomial remains unchanged; only its description varies.   ◇

The passage from vector to coordinates preserves the vector space operations. If v and w have coordinate vectors $[v]_B$ and $[w]_B$ respectively, along with scalar a, then:

1. $[v + w]_B = [v]_B + [w]_B$
2. $[av]_B = a[v]_B$

This preservation of structure means that coordinate vectors themselves form a vector space isomorphic to the original space.

Example 4.8 (Matrix coordinates). Consider the space $\mathbb{R}^{2\times 2}$ with its standard basis of elementary matrices $\mathcal{E} = \{E_{11}, E_{12}, E_{21}, E_{22}\}$. The matrix

$$A = \begin{bmatrix} 2 & -1 \\ 3 & 4 \end{bmatrix}$$

has coordinate vector

$$[A]_{\mathcal{E}} = \begin{pmatrix} 2 \\ -1 \\ 3 \\ 4 \end{pmatrix}$$

relative to this basis. The same matrix written in the alternative basis

$$B = \left\{ \begin{bmatrix} 1 & 0 \\ 0 & 1 \end{bmatrix}, \begin{bmatrix} 0 & 1 \\ 1 & 0 \end{bmatrix}, \begin{bmatrix} 0 & -1 \\ 1 & 0 \end{bmatrix}, \begin{bmatrix} 1 & 0 \\ 0 & -1 \end{bmatrix} \right\}$$

requires different coordinates to express the same transformation.   ◇

The uniqueness of coordinate representations – guaranteed by the linear independence of basis vectors – allows us to test equality through coordinates. Two vectors are equal if and only if their coordinate vectors

Foreshadowing: The preservation of vector space operations under the passage to coordinates explains why matrix multiplication encodes composition of linear transformations.

relative to any basis are equal. This reduces abstract vector equality to numerical comparison.

We have thus established a dictionary between abstract vectors and concrete lists of numbers. This dictionary depends critically on our choice of basis – a choice we are free to make and change as computational needs dictate. The interplay between different choices of basis, and how vectors appear from different coordinate perspectives, leads naturally to our next topic: change of basis transformations.

This reduction of geometric or algebraic properties to numerical tests exemplifies how coordinates enable computation. The art lies in choosing coordinates that make the desired computations simple.

4.3   Change of Basis

The right coordinate system can transform a complex problem into a simple one. An oscillating spring-mass system, described by coupled equations in Cartesian coordinates, reduces to independent motions when viewed in its natural modes. A robotic arm's motion, intricate to specify in workspace coordinates, might follow elementary paths in joint angles. The acceleration of a particle, complicated in rectangular coordinates, could simplify dramatically in polar coordinates. These transformations of perspective – these changes of basis – are not mere mathematical conveniences but essential tools for understanding and controlling physical systems.

Consider a vector space V with two different bases, $\mathcal{B}$ and $\mathcal{B}'$. A vector $v \in V$ exists independently of how we describe it, just as climbing a mountain is equally difficult whether we measure its height in meters or feet. Yet to work with a vector – to compute with it, to transform it, to understand its relationship to other vectors – we must choose coordinates. The same vector has different coordinate representations in different bases:

$$v = \sum_{i=1}^{n} c_i b_i = \sum_{i=1}^{n} c_i' b_i'$$

where $[v]_\mathcal{B} = (c_1, \ldots, c_n)^T$ and $[v]_{\mathcal{B}'} = (c_1', \ldots, c_n')^T$ are its coordinate vectors in bases $\mathcal{B}$ and $\mathcal{B}'$ respectively.

Example 4.9 (Electric Field Components). The electric field $\vec{E}$ from a point charge can be measured in different coordinate systems. Near a charge q at the origin, we might express $\vec{E}$ in Cartesian coordinates:

$$\vec{E} = E_x \hat{\imath} + E_y \hat{\jmath} + E_z \hat{k}$$

or in spherical coordinates:

$$\vec{E} = E_\rho \hat{e}_\rho + E_\theta \hat{e}_\theta + E_\phi \hat{e}_\phi$$

The transformation between these descriptions at any point (x, y, z) with spherical coordinates (ρ, θ, ϕ) is given by:

$$\begin{pmatrix} E_\rho \\ E_\theta \\ E_\phi \end{pmatrix} = \begin{bmatrix} \sin\phi\cos\theta & \sin\phi\sin\theta & \cos\phi \\ -\sin\theta & \cos\theta & 0 \\ \cos\phi\cos\theta & \cos\phi\sin\theta & -\sin\phi \end{bmatrix} \begin{pmatrix} E_x \\ E_y \\ E_z \end{pmatrix}$$

The key to understanding basis changes lies in expressing the new basis vectors in terms of the old, using a matrix to do so.

Definition 4.10 (Change of Basis Matrix). Let $\mathcal{B} = \{b_1, \ldots, b_n\}$ and $\mathcal{B}' = \{b'_1, \ldots, b'_n\}$ be bases for a vector space V. The *change of basis matrix* from $\mathcal{B}$ to $\mathcal{B}'$ is the matrix $P = [p_{ij}]$ whose entries are determined by the unique representations:

$$P = [p_{ij}] \quad : \quad b'_j = \sum_{i=1}^{n} p_{ij} b_i \tag{4.1}$$

The jth column of P contains the $\mathcal{B}$-coordinates of b'_j, encoding how to express each new basis vector in terms of the old basis.

Example 4.11 (Signal Processing). In audio processing, a sound signal naturally begins in the time domain – amplitudes measured at discrete time points. For analysis and filtering, we often transform to the frequency domain using the Discrete Fourier Transform (DFT). This is precisely a change of basis, where our new basis vectors are complex exponentials:

$$b'_k = \frac{1}{\sqrt{n}} \begin{pmatrix} 1 \\ e^{-2\pi i k/n} \\ e^{-4\pi i k/n} \\ \vdots \\ e^{-2\pi i k(n-1)/n} \end{pmatrix}$$

Given the change of basis matrix P, we can convert coordinates systematically:

$$[v]_\mathcal{B} = P[v]_{\mathcal{B}'}$$

This formula reveals that P converts from $\mathcal{B}'$-coordinates to $\mathcal{B}$-coordinates – it "undoes" the basis change. Conversely, to find $\mathcal{B}'$-coordinates given $\mathcal{B}$-coordinates, we solve:

$$[v]_{\mathcal{B}'} = P^{-1}[v]_\mathcal{B}$$

This orthogonal transformation represents a change of basis that proves especially useful in analyzing radially symmetric fields, where the spherical components often reveal patterns obscured in Cartesian coordinates. The transformation matrix follows the mathematicians' convention where θ represents the azimuthal angle in the xy-plane from the x-axis (0 to 2π) and ϕ denotes the polar angle from the z-axis (0 to π).

Foreshadowing: the change of basis matrix P in this case is *unitary* (complex analogues of orthogonal matrices), reflecting the conservation of energy between time and frequency domains.

Example 4.12 (Principal Stress). In analyzing the mechanics of a thin planar material, the *stress tensor* recording stress and strain at a point is a symmetric 2×2 matrix $\sigma \in \text{SYM}_2$ relative to chosen coordinate axes:

$$\sigma = \begin{bmatrix} \sigma_{xx} & \tau_{xy} \\ \tau_{xy} & \sigma_{yy} \end{bmatrix}$$

There always exists a convenient basis – the principal stress directions – in which the stress tensor is diagonal:

$$\sigma' = \begin{bmatrix} \sigma_1 & 0 \\ 0 & \sigma_2 \end{bmatrix}$$

Finding this basis, achieved through eigendecomposition (Chapter 7), is crucial for predicting material failure. The change of basis matrix P here consists of unit vectors along the principal stress directions. $\diamond$

The change of basis matrix P satisfies several properties reflecting the nature of coordinate transformations:

1. P is invertible (else some vectors would lose information in translation)
2. The change from $\mathcal{B}$ to $\mathcal{B}'$ is given by P^{-1} (reversing perspective)
3. For a third basis $\mathcal{B}''$, the matrices compose naturally (reflecting transitivity)

These properties ensure our coordinate transformations are reversible and coherent – changing perspective neither creates nor destroys information about the vectors being described. This preservation of structure lies at the heart of linear algebra's power in engineering: it allows us to work in whatever coordinate system best suits our current needs, confident that we can translate our results back to any other perspective.

The true significance of basis changes emerges when we consider linear transformations, whose matrix representations depend critically on our choice of coordinates. This relationship – between bases, transformations, and their matrix representations – leads us to the fundamental notion of similarity, see Definition 4.16 to follow.

4.4   Matrix Representations

A geometric transformation exists independently of how we measure it: a rotation by 90 degrees clockwise remains the same rotation whether we describe it in Cartesian or polar coordinates. Yet to compute with transformations – to combine them, to apply them to vectors, to analyze their effects – we must express them in coordinates through matrices. The relationship between the abstract transformation and its various matrix representations reveals both the power and limitations of coordinate-based computation.

Let $T : V \to W$ be a linear transformation between vector spaces with chosen bases $\mathcal{B} = \{b_1, \ldots, b_n\}$ for V and $\mathcal{B}' = \{b'_1, \ldots, b'_m\}$ for W. To construct a matrix representation of T, we need only record how it acts on basis vectors:

$$T(b_j) = \sum_{i=1}^{m} a_{ij} b'_i$$

The coefficients a_{ij} form an $m \times n$ matrix $[T]^{\mathcal{B}}_{\mathcal{B}'}$ called the *matrix representation* of T relative to bases $\mathcal{B}$ and $\mathcal{B}'$. The entry a_{ij} gives the ith coordinate of $T(b_j)$ in basis $\mathcal{B}'$.

Think: In $[T]^{\mathcal{B}}_{\mathcal{B}'}$, the bottom basis $\mathcal{B}'$ is where we measure outputs (codomain), while the top basis $\mathcal{B}$ is where we measure inputs (domain). The matrix converts $\mathcal{B}$-coordinates to $\mathcal{B}'$-coordinates, reading from right to left just like function composition.

Example 4.13 (Rotation in Different Bases). Consider the counterclockwise rotation by $\pi/2$ in $\mathbb{R}^2$. In the standard basis $\mathcal{B} = \{\hat{\imath}, \hat{\jmath}\}$, this transformation has the familiar matrix representation:

$$[T]^{\mathcal{B}}_{\mathcal{B}} = \begin{bmatrix} 0 & -1 \\ 1 & 0 \end{bmatrix}$$

Let $\mathcal{B}'$ be the basis consisting of vectors $v_1 = (1,1)^T$ and $v_2 = (-1,2)^T$. The change of basis matrix from $\mathcal{B}'$ to $\mathcal{B}$ is:

$$P = \begin{bmatrix} 1 & -1 \\ 1 & 2 \end{bmatrix}$$

In this new basis, the same rotation transformation appears as:

$$[T]^{\mathcal{B}'}_{\mathcal{B}'} = P^{-1} \begin{bmatrix} 0 & -1 \\ 1 & 0 \end{bmatrix} P = \frac{1}{3} \begin{bmatrix} -1 & -5 \\ 2 & 1 \end{bmatrix}$$

Though the matrices appear quite different, they represent the identical geometric transformation of rotating vectors counterclockwise by $\pi/2$.   ◇

The matrix $[T]^{\mathcal{B}}_{\mathcal{B}'}$ converts input coordinates to output coordinates through standard matrix multiplication:

$$[T(v)]_{\mathcal{B}'} = [T]^{\mathcal{B}}_{\mathcal{B}'} [v]_{\mathcal{B}}$$

This formula encapsulates how linear transformations interact with coordinates: first express the input in $\mathcal{B}$-coordinates, then multiply by the matrix representation to obtain $\mathcal{B}'$-coordinates of the output.

Example 4.14 (Projection onto a Line). Consider the orthogonal projection onto the x-axis in $\mathbb{R}^2$. In standard coordinates, this has matrix representation:

$$[P]^{\mathcal{B}}_{\mathcal{B}} = \begin{bmatrix} 1 & 0 \\ 0 & 0 \end{bmatrix}$$

If we rotate our coordinate system by angle θ, obtaining a new basis $\mathcal{B}' = \{(\cos\theta, \sin\theta)^T, (-\sin\theta, \cos\theta)^T\}$, the same projection appears more complicated:

$$[P]_{\mathcal{B}'}^{\mathcal{B}'} = \begin{bmatrix} \cos^2\theta & \cos\theta\sin\theta \\ \cos\theta\sin\theta & \sin^2\theta \end{bmatrix}$$

The geometric action remains the same – we simply view it through different coordinate lenses. $\diamond$

When bases change, matrix representations transform systematically. If P is the change of basis matrix from $\mathcal{B}'$ to $\mathcal{B}$ in the domain and Q is the change of basis matrix from $\mathcal{D}'$ to $\mathcal{D}$ in the codomain, then:

$$[T]_{\mathcal{D}}^{\mathcal{B}} = Q[T]_{\mathcal{D}'}^{\mathcal{B}'}P^{-1}$$

This relationship reveals how different matrix representations of the same transformation relate through similarity transformations – the subject of our next section.

Example 4.15 (Scaling and Reflection). Consider the linear transformation $T : \mathbb{R}^2 \to \mathbb{R}^2$ that scales by 2 in the x-direction and reflects across the x-axis. In standard coordinates:

$$[T]_{\mathcal{B}}^{\mathcal{B}} = \begin{bmatrix} 2 & 0 \\ 0 & -1 \end{bmatrix}$$

In coordinates rotated by $\pi/4$, this transformation appears to mix scaling and reflection:

$$[T]_{\mathcal{B}'}^{\mathcal{B}'} = \begin{bmatrix} 1/2 & 3/2 \\ 3/2 & 1/2 \end{bmatrix}$$

The apparent complexity arises not from the transformation itself but from our choice of measurement system. $\diamond$

The practical value of matrix representations lies in their computability – they reduce abstract transformations to concrete arrays of numbers. Yet their coordinate dependence reminds us that no single matrix tells the whole story. A transformation may appear simple in one basis and complicated in another, just as a quadratic curve appears differently in different coordinate systems.

Our task in subsequent chapters will be to find bases that reveal the essential features of linear transformations. Some bases will diagonalize certain transformations (Chapter 7); others will respect geometric structure (Chapter 5) or optimize approximations (Chapter 10). Each provides a different lens through which to view and understand linear transformations.

Example: When $\theta = \pi/4$, the basis vectors of $\mathcal{B}'$ are $(\frac{\sqrt{2}}{2}, \frac{\sqrt{2}}{2})^T$ and $(-\frac{\sqrt{2}}{2}, \frac{\sqrt{2}}{2})^T$. The projection matrix in these coordinates becomes

$$[P]_{\mathcal{B}'}^{\mathcal{B}'} = \frac{1}{2}\begin{bmatrix} 1 & 1 \\ 1 & 1 \end{bmatrix}$$

4.5 Coordinate-Free Thought

Our development of coordinates and matrix representations presents a fundamental paradox. We study linear transformations first as abstract mappings between vector spaces, understanding their properties independent of any particular measuring system. Yet to compute with these transformations – to apply them to vectors, to compose them, to analyze their effects – we must choose coordinates and work with matrices. This tension between the intrinsic nature of transformations and their coordinate-dependent representations lies at the heart of linear algebra.

Foreshadowing: The relationship between isomorphic vector spaces and similar transformations hints at deeper categorical structures in mathematics.

Consider data drawn from some high-dimensional measurement process – perhaps gene expression levels across thousands of cells, or activation patterns across layers of a neural network. The underlying biological or computational structure exists independent of how we choose to measure it. Different experimental protocols or network architectures may yield different representations of the same fundamental patterns. This suggests a deeper question: when are two apparently different representations truly equivalent?

Some find the equivalent expression

$$AX = XB$$

to be more memorable and evocative.

Definition 4.16 (Similarity). Two matrices A and B are *similar* if there exists an invertible matrix X such that:

$$B = X^{-1}AX$$

We write $A \sim B$ to denote similar matrices.

This algebraic relationship captures precisely when two matrices represent the same linear transformation viewed through different coordinate systems. The matrix P encodes the change of basis that transforms one view into another. More fundamentally, we say two linear transformations $S, T : V \to V$ are *similar* if there exists an isomorphism $\varphi : V \to V$ such that $S = \varphi^{-1}T\varphi$. When expressed in coordinates, this abstract notion manifests as matrix similarity.

The term *conjugate* is more common in mathematics, but *similar* will do nicely.

Example 4.17 (Geometric Similarity). Consider a rotation by 90° counterclockwise in the plane. In standard coordinates, this appears as:

$$A = \begin{bmatrix} 0 & -1 \\ 1 & 0 \end{bmatrix}$$

This diagram commutes when $S = \varphi^{-1}T\varphi$, illustrating similarity as conjugation by the isomorphism φ.

If we measure vectors instead using the basis $\{v_1, v_2\}$ where:

$$v_1 = \begin{pmatrix} 1 \\ 1 \end{pmatrix}, \quad v_2 = \begin{pmatrix} -1 \\ 2 \end{pmatrix} \quad \Rightarrow \quad X = \begin{bmatrix} 1 & -1 \\ 1 & 2 \end{bmatrix}$$

then the same rotation has matrix:

$$B = X^{-1}AX = \begin{bmatrix} 1/3 & -1 \\ 1/3 & 0 \end{bmatrix}$$

Though these matrices look quite different, they represent identical geometric actions. ◇

Example 4.18 (Shape Analysis). Consider analyzing the shape of a complex three-dimensional object represented as a cloud of points in $\mathbb{R}^3$. The intrinsic shape exists independent of coordinate choice – rotating or translating the object does not change its fundamental geometry. Two different coordinate representations of the same shape are related through similarity transformations. Geometric approaches to data analysis build on this insight, studying properties of shapes that remain invariant under such transformations. ◇

Similar matrices share certain properties that are intrinsic to the transformation they represent:

1. They have the same determinant
2. They have the same rank
3. They have the same trace (sum of diagonal entries)

These *coordinate invariants* belong to the transformation itself rather than to any particular matrix representation. They form a fingerprint that distinguishes genuinely different transformations from merely different coordinate views of the same map.

The distinction between intrinsic and coordinate-dependent properties guides much of modern data analysis. When studying high-dimensional data, we ask:

1. What features are invariant under reasonable transformations?
2. Which coordinate systems best reveal these features?
3. How do different representations illuminate different aspects?

The power of this perspective emerges in the chapters ahead. We will discover:

- Coordinates that respect geometric structure (Chapter 5)
- Representations that reveal dynamical behavior (Chapter 7)
- Bases optimal for dimension reduction (Chapter 10)

Each provides a different lens through which to view our objects of study, highlighting different features while preserving their essential character.

Example 4.19 (Data Visualization). A high-dimensional dataset can be visualized in countless ways through different projections onto two or three

Foreshadowing: In Chapter 7, we will discover that similar matrices also share eigenvalues – another intrinsic property of the transformation they represent.

Foreshadowing: In Chapter 11, we will see how Principal Component Analysis discovers coordinate systems that reveal intrinsic low-dimensional structure in high-dimensional data.

dimensions. While some projections may obscure the data's structure, others reveal meaningful patterns. The art lies in finding coordinates that expose the features of interest while respecting the intrinsic relationships in the data. Though the choice of visualization coordinates is arbitrary, the patterns they reveal – clusters, outliers, nonlinear relationships – reflect genuine structure in the data. ◇

Yet we must remain mindful of the difference between a transformation and its various representations. Matrices are tools for computation – powerful and necessary tools, but not the whole story. The true objects of study are the transformations themselves, existing independent of how we choose to measure them. This tension between coordinate-free concepts and coordinate-based calculations runs throughout linear algebra, from the most theoretical ideas to the most practical applications.

Our journey through linear algebra will constantly navigate between these perspectives – between the abstract and the concrete, between the geometric and the algebraic, between the intrinsic and the coordinate-dependent. The art lies in knowing when to reason coordinate-free and when to harness the computational power of well-chosen coordinates. In the chapters ahead, we will see this interplay illuminate everything from the structure of differential equations to the architecture of neural networks.

Robotic Arm Kinematics

The mathematics of robotic manipulation provides a compelling demonstration of how different coordinate systems illuminate different aspects of the same physical system. A robotic arm's motion can be described through multiple bases, each revealing different aspects of its behavior. These coordinate choices — and the transformations between them — exemplify the fundamental principles developed throughout this chapter.

Consider a planar robotic arm with two revolute joints connecting two rigid links of lengths L_1 and L_2. The configuration of this arm admits two natural coordinate systems:

1. Joint space coordinates (θ_1, θ_2), measuring the angles of each joint
2. Task space coordinates (x, y), giving the position of the end-effector

These spaces come equipped with natural bases: joint space has basis vectors corresponding to infinitesimal rotations of each joint, while task space inherits the standard Cartesian basis of the plane. The transformation between these coordinates is given by:

$$F(\theta_1, \theta_2) = \begin{pmatrix} L_1 \cos\theta_1 + L_2 \cos(\theta_1 + \theta_2) \\ L_1 \sin\theta_1 + L_2 \sin(\theta_1 + \theta_2) \end{pmatrix}$$

Though this transformation F is nonlinear, its derivative $[DF]$ at any configuration provides a linear map between the tangent spaces — a change of basis matrix relating infinitesimal motions:

$$\begin{pmatrix} \dot{x} \\ \dot{y} \end{pmatrix} = [DF]_{(\theta_1,\theta_2)} \begin{pmatrix} \dot{\theta}_1 \\ \dot{\theta}_2 \end{pmatrix}$$

where:

$$[DF]_{(\theta_1,\theta_2)} = \begin{bmatrix} -L_1 \sin\theta_1 - L_2 \sin(\theta_1 + \theta_2) & -L_2 \sin(\theta_1 + \theta_2) \\ L_1 \cos\theta_1 + L_2 \cos(\theta_1 + \theta_2) & L_2 \cos(\theta_1 + \theta_2) \end{bmatrix}$$

The columns of this matrix express the task-space velocities generated by unit joint velocities — they form a configuration-dependent basis for achievable end-effector motions.

More complex manipulators illuminate additional aspects of coordinate relationships. Consider extending our arm to three joints while maintaining planar end-effector motion. Now joint space has basis vectors $\{\partial/\partial\theta_1, \partial/\partial\theta_2, \partial/\partial\theta_3\}$ while task space remains two-dimensional with basis $\{\partial/\partial x, \partial/\partial y\}$. The derivative becomes a linear transformation $[DF] : \mathbb{R}^3 \to \mathbb{R}^2$ between these spaces, with:

$$[DF]_{(\theta_1,\theta_2,\theta_3)} = \begin{bmatrix} \dfrac{\partial x}{\partial\theta_1} & \dfrac{\partial x}{\partial\theta_2} & \dfrac{\partial x}{\partial\theta_3} \\ \dfrac{\partial y}{\partial\theta_1} & \dfrac{\partial y}{\partial\theta_2} & \dfrac{\partial y}{\partial\theta_3} \end{bmatrix}$$

Recall: The Inverse Function Theorem from calculus guarantees that when $[DF]$ is invertible at a configuration (i.e., when $\det[DF] \neq 0$), F has a local inverse — we can solve uniquely for small changes in joint angles needed to achieve desired end-effector motions. When $[DF]$ fails to be invertible, as happens when the arm is fully extended, certain instantaneous motions become impossible.

This matrix has a nontrivial kernel — reflecting joint velocities that instantaneously leave the end-effector fixed. Such self-motions exemplify the fundamental kernel-image relationship studied in Chapter 3, now emerging naturally in a concrete mechanical system.

The practical significance of these coordinate relationships manifests in trajectory planning. A straight-line motion of the end-effector, though elegant in task coordinates, may demand intricate joint-space choreography. Conversely, simple joint trajectories can trace complex paths through task space. For the three-joint arm, redundancy enriches this relationship further — the same end-effector trajectory admits infinitely many joint space realizations, corresponding to different paths through the kernel of $[DF]$.

This duality between representations reflects a deeper truth: no single coordinate system captures all aspects of a complex system with equal clarity. The art of engineering lies not merely in choosing appropriate coordinates but in moving fluently between different representations as the problem demands. Joint coordinates render questions of dynamics and joint limits transparent, while task coordinates simplify motion specification. The mathematical framework developed in this chapter transforms this art from intuitive craft to systematic science, providing the tools needed to work effectively with multiple coordinate representations of the same underlying reality.

Computer Graphics & Coordinate Systems

Modern computer graphics illuminates the practical power of coordinate transformations. A virtual object – perhaps a spacecraft in a flight simulator – exists simultaneously in multiple coordinate systems, each chosen to simplify particular aspects of the simulation. Understanding how these bases relate through the transformations of Section 4.3 converts complex geometric problems into systematic matrix computations.

Consider our virtual spacecraft. Its geometry begins life in *body coordinates,* where the natural basis $\mathcal{B}_b = \{b_1, b_2, b_3\}$ aligns with the craft's structure: b_1 points through the nose, b_2 along the right wing, and b_3 downward. In these coordinates, the craft's symmetries become apparent and control surfaces align with coordinate planes. A point p on the spacecraft has coordinate vector $[p]_{\mathcal{B}_b}$ relative to this body basis.

Yet our spacecraft moves through a virtual world with its own coordinate system. The *world basis* $\mathcal{B}_w = \{w_1, w_2, w_3\}$ typically aligns w_3 with vertical, while w_1 and w_2 span the ground plane. Following Section 4.2, the coordinate transformation from body to world basis follows from the change of basis matrix:

$$[p]_{\mathcal{B}_w} = P[p]_{\mathcal{B}_b}$$

This transformation matrix P has columns expressing body basis vectors in world coordinates, exactly as constructed in Example 7.3:

$$P = \begin{bmatrix} | & | & | \\ [b_1]_{\mathcal{B}_w} & [b_2]_{\mathcal{B}_w} & [b_3]_{\mathcal{B}_w} \\ | & | & | \end{bmatrix}$$

Each column shows how one body basis vector decomposes in world coordinates. Like the rotation matrices studied in Section 3.1, this change of basis preserves lengths and angles – a crucial property for rigid objects.

A virtual camera introduces yet another basis. The *camera basis* $\mathcal{B}_c = \{c_1, c_2, c_3\}$ places the virtual lens at the origin with c_3 pointing along the viewing direction and c_2 aligned with the image's vertical axis. Points transform to these coordinates through composition with another change of basis matrix Q:

$$[p]_{\mathcal{B}_c} = Q[p]_{\mathcal{B}_w}$$

The composition of these transformations – from body to world to camera coordinates – embodies the core algebraic insight of Section 4.3: changes of basis compose through matrix multiplication. A point's coordinates transform as:

$$[p]_{\mathcal{B}_c} = QP[p]_{\mathcal{B}_b}$$

This matrix product captures the complete change of coordinates, though we often maintain separate transformations for clarity and efficiency.

Each basis in this sequence serves a specific purpose: body coordinates for physics simulation, world coordinates for scene composition, camera coordinates for visibility and rendering. The transformations between them, though apparently complex, follow directly from our precise understanding of coordinates and

Example: When a spacecraft pitches upward 30°, its body basis vectors expressed in world coordinates become columns of the change of basis matrix P:

$$\begin{bmatrix} \sqrt{3}/2 & 0 & -1/2 \\ 0 & 1 & 0 \\ 1/2 & 0 & \sqrt{3}/2 \end{bmatrix}$$

These columns are mutually orthogonal since P represents rigid rotation.

bases developed in Section 4.2. This exemplifies a broader principle: challenging problems often become tractable when viewed in appropriate coordinates.

The practical significance extends far beyond graphics. In robotics, similar coordinate changes relate joint angles to end-effector position through the transformation matrices studied in Section 4.4. In computer vision, camera and world bases must align to enable augmented reality. In spacecraft guidance, body and inertial coordinates interplay in navigation algorithms. Each application builds on the same mathematical foundation: the careful construction of bases and transformations between them.

This cascade of coordinate systems illustrates a final key insight from Section 4.5: bases should be chosen to match the natural structure of our problems. Body coordinates respect vehicle symmetries; world coordinates align with gravity and terrain; camera coordinates match viewing geometry. No single basis captures all aspects elegantly – the art lies in choosing appropriate bases for each subtask while understanding how they relate through the careful matrix algebra developed in this chapter.

Exercises: Chapter 4

1. For the basis $B = \{v_1, v_2\}$ of $\mathbb{R}^2$ where $v_1 = (1,2)^T$ and $v_2 = (1,-1)^T$, find the change of basis matrix P from B to the standard basis.

2. Consider $v = (2,1,-1)^T$ in $\mathbb{R}^3$ and basis $B = \{(1,1,0)^T, (0,1,1)^T, (1,0,1)^T\}$. Find $[v]_B$.

3. Find a basis for the space SYM_2 of symmetric 2×2 matrices, then find the coordinate vector of $\begin{bmatrix} 3 & 1 \\ 1 & 2 \end{bmatrix}$ in your basis.

4. Find the coordinate vector of $p(x) = x^2 - 2x + 1$ relative to the basis $\{1 + x^2, x - x^2, 1 - x\}$ of $\mathcal{P}_2$.

5. For $\mathcal{P}_2$, show that $\{1, 1+x, 1+x+x^2\}$ is a basis by proving it is linearly independent and spans $\mathcal{P}_2$. Then express $\{1, x, x^2\}$ in terms of this basis.

6. Given a basis $B = \{v_1, v_2\}$ for $\mathbb{R}^2$, prove that the change of basis matrix from B to the standard basis has columns equal to v_1 and v_2.

7. Let V be the space of 2×2 matrices with basis $B = \{E, S, R, D\}$ where $E = \begin{bmatrix} 1 & 0 \\ 0 & 1 \end{bmatrix}$, $S = \begin{bmatrix} 0 & 1 \\ 1 & 0 \end{bmatrix}$, $R = \begin{bmatrix} 0 & -1 \\ 1 & 0 \end{bmatrix}$, $D = \begin{bmatrix} 1 & 0 \\ 0 & -1 \end{bmatrix}$. Describe a method for finding the coordinate vector of any 2×2 matrix in this basis.

8. Let $T : \mathbb{R}^2 \to \mathbb{R}^2$ rotate vectors counterclockwise by $\pi/4$. Find its matrix in: (a) the standard basis, (b) the basis $\{(1,1)^T, (-1,1)^T\}$.

9. Let $T : \mathcal{P}_2 \to \mathcal{P}_2$ be differentiation $T(p) = p'$. Find its matrix in the standard basis $\{1, x, x^2\}$, then find a basis giving it a simpler matrix representation.

10. Consider matrices $A = \begin{bmatrix} 1 & 2 \\ 3 & 4 \end{bmatrix}$ and $B = \begin{bmatrix} 5 & -2 \\ -3 & 0 \end{bmatrix}$. Either find a matrix P such that $B = P^{-1}AP$, or prove no such P exists. What invariants could you check first?

11. For $T : \mathbb{R}^2 \to \mathbb{R}^2$ with matrix $[T]_{B_1} = \begin{bmatrix} 2 & 1 \\ 1 & 2 \end{bmatrix}$ in basis B_1 and $[T]_{B_2} = \begin{bmatrix} 3 & 0 \\ 0 & 1 \end{bmatrix}$ in basis B_2, find the change of basis matrix from B_2 to B_1.

12. The matrices $A = \begin{bmatrix} 3 & 1 \\ 1 & 3 \end{bmatrix}$ and $B = \begin{bmatrix} 2 & 2 \\ 2 & 4 \end{bmatrix}$ are similar. How can one prove this directly? Try finding P such that $B = P^{-1}AP$. Can this be done by solving a linear system?

13. Consider the operator $T : \mathcal{P}_2 \to \mathcal{P}_2$ defined by $T(p)(x) = p(x+1)$. Find its matrix in the standard basis $\{1, x, x^2\}$. Then find its matrix in basis $\{1, x + 1, (x+1)^2\}$. What pattern do you observe? Explain why this pattern occurs geometrically.

14. Consider the basis $\{v_1, v_2, v_3\}$ for $\mathbb{R}^3$ where $v_1 = (1,1,1)^T$, $v_2 = (1,1,-2)^T$, $v_3 = (1,-2,1)^T$. For $T : \mathbb{R}^3 \to \mathbb{R}^3$ given by the cross product $T(x) = x \times (1,0,0)^T$, find the matrix of T in this basis. First find how T acts on each basis vector, then combine these into the matrix representation.

15. For a linear transformation $T : V \to V$: (a) if $[T]_B^B$ is diagonal for some basis B, explain geometrically what this means about how T acts on vectors, (b) prove that finding such a basis is equivalent to finding vectors that T maps to scalar multiples of themselves.

16. For similar matrices A and B: (a) prove $\det(A) = \det(B)$ and $\text{tr}(A) = \text{tr}(B)$; (b) show also that A^n and B^n are similar for $n > 1$. *Hint:* you know how determinants behave under composition; for trace, recall that (or show) $\text{tr}(XY) = \text{tr}(YX)$.

17. For a 2×2 matrix A, prove that if A is similar to a diagonal matrix, then it is similar to any matrix with the same diagonal entries (in any order). Give an example showing two matrices with the same diagonal entries need not be similar.

18. Let $\alpha(t) = 1 + t + t^2$ and $\beta(t) = 1 - t^2$ be polynomials in $\mathcal{P}_2$. Find bases B_1 and B_2 such that $[\alpha]_{B_1} = (1,0,0)^T$ and $[\beta]_{B_2} = (1,0,0)^T$. What does this tell you about these polynomials? Find the change of basis matrix from B_1 to B_2.

19. For a vector space V with basis $B = \{v_1, \dots, v_n\}$, prove that any linear transformation $T : V \to V$ is uniquely determined by its action on basis vectors. Then show this implies similar matrices must have the same trace without using matrix multiplication. How does this connect to the geometry of linear transformations?

Chapter 5
Inner Products & Orthogonality

"others triangular right angled course maintain; others obtuse, acute Scalene, in simple paths"

THE SPACES WE INHABIT POSSESS structure beyond mere addition and scaling. Like a compass measuring celestial arcs, vectors trace paths that stretch and bend through space – each with its own magnitude, each forming precise angles with its fellows. These geometric notions – of length and perpendicularity, of measure and relation – emerge not from arbitrary convention but from careful definition of how vectors interact through inner products.

Our task is to extract the essence of geometric measurement – of lengths, angles, and orthogonality – and extend it beyond its Euclidean origins. The familiar dot product of vectors served well in calculus, yet it represents merely one instance of a deeper structure. Inner products provide the machinery to impose geometric order on abstract vector spaces, enabling us to measure and compare vectors in ways that respect their intrinsic nature.

This geometric perspective transforms our understanding of linear algebra. Orthogonal vectors, previously understood through coordinate calculations, emerge as a fundamental organizing principle. Orthogonal bases offer optimal frameworks for computation. Orthogonal matrices preserve the geometric structure we construct. Through inner products, the abstract vector spaces of previous chapters acquire shape and substance.

The choice of inner product shapes our view of a vector space, highlighting certain features while obscuring others. Different inner products induce different notions of length and angle, each suited to particular applications. Some arise naturally from physical principles, others from statistical considerations, still others from computational convenience.

The art lies in choosing an inner product that illuminates the aspects of greatest interest while preserving essential structure.

Our development proceeds from the concrete to the abstract and back again. The familiar dot product guides our intuition as we ascend axiomatically. Though we shall occasionally glimpse infinite-dimensional spaces through carefully chosen examples, our focus remains on finite-dimensional spaces where the theory achieves its purest form. Each definition builds upon the last with mathematical certainty, yet the structure we construct reflects something fundamental: the deep unity between geometric intuition and algebraic law.

5.1   Dot & Inner Products

The dot product pushes out from its first appearance in calculus. Two vectors $u, v \in \mathbb{R}^n$ combine through coordinate-wise multiplication and addition:

$$u \cdot v = \sum_{i=1}^{n} u_i v_i$$

This operation, though defined through coordinates, reveals fundamental geometric features: length through $\|v\| = \sqrt{v \cdot v}$, angle via $u \cdot v = \|u\|\|v\|\cos\theta$, and orthogonality when $u \cdot v = 0$. That such a simple formula encodes so much geometric content suggests deeper structure at play.

Consider what properties make the dot product geometrically meaningful. First, it treats vectors symmetrically: $u \cdot v = v \cdot u$. Second, it is linear in each factor: $(cu + w) \cdot v = c(u \cdot v) + w \cdot v$. Third, it ensures positive length: $v \cdot v \geq 0$, with equality only when $v = 0$. These properties – not the specific formula – enable geometric measurement.

This insight suggests generalizing beyond $\mathbb{R}^n$. Consider the space $C([0,1])$ of continuous functions on the unit interval. Though these vectors are curves rather than arrows, we might still wish to measure angles between them or test their orthogonality. The integral formula

$$\langle f, g \rangle = \int_0^1 f(t)g(t)\, dt$$

provides exactly such a measurement. It shares the key properties that made the dot product geometric: symmetry, linearity, and positivity. Two functions are now "orthogonal" when their product integrates to zero – a concept familiar from calculus (and, later, Fourier analysis).

These examples motivate the abstract definition that captures their common essence:

Example: The functions $\sin(n\pi x)$ and $\sin(m\pi x)$ are orthogonal for distinct integers n, m under this inner product, explaining the independence of Fourier sine series terms.

Definition 5.1 (Inner Product). An *inner product* on a vector space V is a function $\langle \cdot, \cdot \rangle : V \times V \to \mathbb{R}$ satisfying, for all $u, v, w \in V$ and $c \in \mathbb{R}$:

1. Symmetry: $\langle u, v \rangle = \langle v, u \rangle$
2. Linearity: $\langle cu + w, v \rangle = c\langle u, v \rangle + \langle w, v \rangle$
3. Positive Definiteness: $\langle v, v \rangle \geq 0$, with equality if and only if $v = 0$

A vector space equipped with an inner product is called an *inner product space*.

This seemingly austere definition distills the essential features that enable geometric measurement. Each property plays a vital role: symmetry ensures angles are well-defined; linearity connects geometry to vector space structure; positive definiteness guarantees meaningful notions of length and distance.

Example 5.2 (Weighted Inner Products). On $\mathbb{R}^n$, we need not weight all coordinates equally. Given positive weights $a_1, \ldots, a_n$, the formula

$$\langle u, v \rangle_a = \sum_{i=1}^{n} a_i u_i v_i$$

defines an inner product that emphasizes certain components over others. Such weighted measurements arise naturally in statistics, where the weights might reflect measurement uncertainty, or in mechanics, where they encode mass distribution. $\diamond$

Example 5.3 (Matrix Inner Products). The space $\mathbb{R}^{m \times n}$ of matrices admits several natural inner products. The *Frobenius inner product*,

$$\langle A, B \rangle_F = \mathrm{tr}(A^T B) = \sum_{i,j} a_{ij} b_{ij}$$

treats a matrix as a long vector of entries. Other inner products might weight different matrix entries according to their positions or statistical significance. $\diamond$

Each inner product imposes its own geometry on a vector space, determining how angles and lengths are measured. The dot product is merely first among equals – the most elementary inner product on $\mathbb{R}^n$. As we shall see, the choice of inner product profoundly affects both theoretical understanding and practical computation.

The most fundamental consequence of an inner product is its induced notion of length.

Foreshadowing: The choice of inner product shapes everything from optimization algorithms to quantum measurements. We shall see its influence grow throughout this text.

Definition 5.4 (Norm). Given an inner product space V, the *norm* of a vector $v \in V$ is defined by:

$$\|v\| = \sqrt{\langle v, v \rangle}$$

This induced norm measures the length of vectors in a way compatible with the inner product structure.

Though many norms exist on vector spaces, those arising from inner products possess special geometric properties. The subtle interplay between inner products and their induced norms will guide our development of orthogonality in the sections ahead.

5.2   Angles & Orthogonality

In physical space, vectors meet at angles. This seemingly elementary observation – that two directions can be more or less aligned – extends far beyond geometry. Two functions can be more or less correlated; two matrices can be more or less aligned; two quantum states can be more or less independent. In each case, the inner product reveals this angular relationship through a profound formula first glimpsed in calculus.

Our development requires first a fundamental inequality – one that ensures angles make sense in any inner product space.

Lemma 5.5 (Cauchy-Schwarz Inequality). *For any vectors u, v in an inner product space,*

$$|\langle u, v \rangle| \leq \|u\| \|v\|$$

with equality if and only if one vector is a scalar multiple of the other.

Proof. For any real number t, positive-definiteness of the inner product requires:

$$0 \leq \|u + tv\|^2 = \|u\|^2 + 2t\langle u, v \rangle + t^2\|v\|^2$$

This quadratic in t must be nonnegative for all t, possible only if its discriminant is nonpositive:

$$4\langle u, v \rangle^2 \leq 4\|u\|^2\|v\|^2$$

The case of equality follows by examining when this quadratic has exactly one root. $\square$

This seemingly technical result has profound implications. It ensures that the ratio of inner product to product of norms cannot exceed unity in absolute value – exactly what we need to define angles through the familiar cosine relationship.

The *angle* between nonzero vectors u and v in an inner product space is the unique number $\theta \in [0, \pi]$ satisfying:

$$\cos \theta = \frac{\langle u, v \rangle}{\|u\| \|v\|} \tag{5.1}$$

When this angle is $\pi/2$, we say the vectors are *orthogonal* and write $u \perp v$.

That angles exist in all inner product spaces is remarkable. More remarkable still is how the familiar properties of Euclidean angles persist in the abstract setting. Orthogonal vectors have inner product zero; small angles correspond to nearly parallel vectors; obtuse angles arise when the inner product is negative. These properties transcend the specific setting, whether we measure angles between functions, matrices, or quantum states.

Example 5.6 (Function Angles). In $C([a,b])$ with the inner product $\langle f, g \rangle = \int_a^b f(t)g(t)\, dt$, two functions are orthogonal when their product integrates to zero. The angle between functions measures their correlation – how their variations align over the interval. Functions with small angle vary similarly; orthogonal functions vary independently; functions at angle π vary in opposition.   ◇

Example: The functions $\sin x$ and $\cos x$ are orthogonal on $[-\pi, \pi]$ under this inner product – a fact crucial to Fourier analysis. See Example 5.9.

Example 5.7 (Matrix Alignment). Under the Frobenius inner product, matrices $A, B \in \mathbb{R}^{m \times n}$ form an angle through:

$$\cos \theta = \frac{\operatorname{tr}(A^T B)}{\|A\|_F \|B\|_F}$$

This measures how aligned their entries are, with orthogonal matrices having zero entry-wise correlation. Such geometric interpretation of matrix relationships reveals structure hidden in algebraic formulas.   ◇

Orthogonality proves especially powerful in decomposing vectors. When $u \perp v$, the Pythagorean theorem generalizes:

$$\|u + v\|^2 = \|u\|^2 + \|v\|^2$$

This additivity of squared norms for orthogonal vectors enables decomposition of complex vectors into simpler orthogonal components – a principle that will guide our study of orthogonal bases and projections.

Lemma 5.8 (Orthogonal Decomposition). *Let $v_1, \ldots, v_k$ be mutually orthogonal nonzero vectors. Then they are linearly independent, and for any scalars $c_1, \ldots, c_k$:*

$$\left\| \sum_{i=1}^k c_i v_i \right\|^2 = \sum_{i=1}^k c_i^2 \|v_i\|^2$$

Foreshadowing: This decomposition principle will reach its full power when we construct orthonormal bases, enabling optimal approximations and computational methods.

The proof follows from the distributive property of inner products and the vanishing of cross terms between orthogonal vectors. More significant is the implication: orthogonal vectors combine independently, their contributions to any sum measurable separately without interference. This independence principle – that orthogonal components can be analyzed separately – pervades modern applications from signal processing to quantum mechanics.

We close with a subtle observation: while every inner product induces a norm, not every norm arises from an inner product. The ℓ^1 and ℓ^∞ norms on $\mathbb{R}^n$, for instance, lack the geometric structure that inner products provide. The special character of inner product norms lies in how they encode angles – a capability we shall exploit as we develop the theory of orthogonal bases and transformations.

5.3   Orthogonal & Orthonormal Bases

The bases we have thus far encountered arose from convenience or custom – coordinates chosen more by habit than principle. Yet some bases are intrinsically better than others, measuring vectors in ways that respect the inner product structure we have so carefully constructed. Such bases emerge from the concept of orthogonality, providing optimal frameworks for both theoretical understanding and practical computation.

A set of vectors $\{v_1, \ldots, v_k\}$ in an inner product space is *orthogonal* if each vector is perpendicular to all others:

$$\langle v_i, v_j \rangle = 0 \quad \text{for all } i \neq j$$

When these vectors also have unit length, so that $\|v_i\| = 1$ for all i, we call the set *orthonormal*. Such collections combine the geometric elegance of perpendicularity with the computational convenience of unit vectors.

Example: The standard basis $\{\hat{\imath}, \hat{\jmath}, \hat{k}\}$ for $\mathbb{R}^3$ is orthonormal under the dot product – a fact so familiar we often forget its significance.

The power of orthogonal vectors lies in how they decompose the space they span. When $v_1, \ldots, v_k$ are orthogonal, any vector in their span has a unique representation:

$$x = \sum_{i=1}^{k} c_i v_i \quad \text{where} \quad c_i = \frac{\langle x, v_i \rangle}{\|v_i\|^2}$$

The coefficients emerge naturally from the inner product – no system of equations need be solved. When the vectors are orthonormal, this simplifies further to $c_i = \langle x, v_i \rangle$, the inner product itself revealing the coordinates directly.

Example 5.9 (Fourier Series). The functions $\{\sin nx, \cos nx\}_{n=1}^{\infty}$ form an orthogonal set in $C[-\pi, \pi]$ under the inner product

$$\langle f, g \rangle = \int_{-\pi}^{\pi} f(x) g(x) \, dx$$

This orthogonality – discovered by Euler and exploited by Fourier – explains why trigonometric series decompose periodic functions so effectively. The Fourier coefficients arise naturally as inner products, without need for integration by parts or other technical devices.   ◇

An orthogonal or orthonormal set that spans a space forms a basis of particular elegance. Every vector has unique coordinates computable through inner products; the Pythagorean theorem holds for all linear combinations; geometric and algebraic properties align perfectly. Yet we cannot simply wish such bases into existence – we must construct them systematically.

The *Gram-Schmidt process* provides such construction. Beginning with any basis $\{b_1, \ldots, b_n\}$, we build an orthogonal basis $\{v_1, \ldots, v_n\}$ spanning the same space:

$$
\begin{aligned}
v_1 &= b_1 \\
v_2 &= b_2 - \Pi_{v_1} b_2 \\
v_3 &= b_3 - \Pi_{v_1} b_3 - \Pi_{v_2} b_3 \\
&\ \ \vdots \\
v_k &= b_k - \sum_{i=1}^{k-1} \Pi_{v_i} b_k
\end{aligned}
$$

where $\Pi_v u = \frac{\langle u,v \rangle}{\|v\|^2} v$ denotes orthogonal projection. Each new vector is made orthogonal to all previous ones by subtracting away its components in their directions.

Perspective: While Gram-Schmidt orthogonalization may seem purely theoretical, it underlies key algorithms in modern recommender systems, where orthogonal feature vectors help prevent redundancy in user representations.

Example 5.10 (Polynomial Orthogonalization). Consider the space $\mathcal{P}_4$ with inner product $\langle f, g \rangle = \int_{-1}^{1} f(x)g(x)\,dx$. Starting with the monomial basis $\{1, x, x^2, x^3, x^4\}$, Gram-Schmidt produces (up to scaling) the *Legendre polynomials*:

$$
\begin{aligned}
P_0(x) &= 1 \\
P_1(x) &= x \\
P_2(x) &= \tfrac{1}{2}(3x^2 - 1) \\
P_3(x) &= \tfrac{1}{2}(5x^3 - 3x) \\
P_4(x) &= \tfrac{1}{8}(35x^4 - 30x^2 + 3)
\end{aligned}
$$

Remarkably, these polynomials play a fundamental role not only in gravitational theory but also in quantum mechanics, where they describe angular momentum states, and in numerical integration, where they provide optimal quadrature points.

These orthogonal polynomials, discovered by Legendre in studying gravitational potential, arise naturally from imposing orthogonality on the simplest polynomial basis. The increasing complexity of coefficients reflects how each new polynomial must maintain orthogonality to all previous ones – a constraint leading to ever more intricate balancing of terms.   ◇

The Gram-Schmidt process, though elegant in theory, can suffer numerical instability in practice. Each projection accumulates computational errors that can destroy orthogonality in the final basis. A more stable approach – *modified Gram-Schmidt* – applies the projections sequentially rather than simultaneously. Though mathematically equivalent, this version better preserves orthogonality in finite-precision arithmetic.

Having constructed orthogonal bases, we might ask which are best suited to particular problems. The answer depends on what structure we wish to preserve or illuminate:

1. For differential equations, bases of eigenfunctions reveal dynamical behavior
2. In signal processing, Fourier bases expose frequency content
3. For quantum systems, bases of energy eigenstates clarify measurement
4. In data analysis, principal component bases optimize variance capture

The choice of orthogonal basis shapes our view of the space and its vectors – a theme we shall explore more deeply when studying eigenvalues and singular values.

5.4   Adjoints & Transposes

The familiar operation of matrix transpose harbors deeper structure than first appears. When we write A^T for a matrix A, we do more than reflect entries across the diagonal – we encode a fundamental relationship between linear transformations and inner products. This relationship, abstracted from its matrix origins, provides the key to understanding how transformations interact with geometric structure.

Consider first the matrix transpose in $\mathbb{R}^n$ with its standard inner product. For any matrix A, its transpose A^T satisfies a crucial property: for all vectors x and y,

$$\langle Ax, y \rangle = \langle x, A^T y \rangle$$

This seemingly innocent equation reveals something profound: the transpose A^T is not merely a matrix operation but rather the unique linear transformation that preserves inner product relationships with A.

Definition 5.11 (Adjoint). Let V and W be inner product spaces and $T : V \to W$ a linear transformation. The *adjoint* of T is the unique linear transformation $T^* : W \to V$ satisfying:

$$\langle Tv, w \rangle_W = \langle v, T^* w \rangle_V \tag{5.2}$$

for all $v \in V$ and $w \in W$. ●

Example 5.12 (Differentiation Adjoint). Consider the differentiation operator $D : \mathcal{P}_2 \to \mathcal{P}_1$ with inner product $\langle f, g \rangle = \int_0^1 f(x)g(x)\,dx$. Its adjoint

$D^* : \mathcal{P}_1 \rightarrow \mathcal{P}_2$ satisfies:

$$\int_0^1 (Df)(x)g(x)\,dx = \int_0^1 f(x)(D^*g)(x)\,dx$$

Integration by parts reveals that D^* involves both integration and boundary terms – a relationship fundamental to both differential equations and variational principles in mechanics. ◇

The adjoint operation respects the algebraic structure of linear transformations while reversing their direction:

Nota bene: the theory of adjoints extends to infinite-dimensional spaces, but requires additional machinery from functional analysis including completeness and boundedness of operators.

Lemma 5.13. *For linear transformations S and T between finite-dimensional inner product spaces, and for any scalar c:*

1. $(S+T)^* = S^* + T^*$
2. $(cT)^* = cT^*$
3. $(ST)^* = T^*S^*$
4. $(T^*)^* = T$

When we choose orthonormal bases for our spaces, the matrix of T^* is the transpose of the matrix of T. This explains our notation: the abstract adjoint operation generalizes matrix transpose to arbitrary inner product spaces. The matrix transpose is merely the adjoint operation expressed in coordinates.

Example: For a rotation matrix R in $\mathbb{R}^2$, the adjoint R^* corresponds to rotation in the opposite direction – explaining why $R^T R = I$.

Example 5.14 (Signal Processing). In digital signal processing, filtering operations are linear transformations between signal spaces. The adjoint of a filter describes its time-reversed impulse response – a relationship crucial to matched filtering and signal detection. When the filter preserves signal energy (an inner product constraint), its adjoint is closely related to its inverse. ◇

The relationship between a transformation and its adjoint illuminates the geometry of linear mappings. Some transformations equal their own adjoints ($T = T^*$); others satisfy $T^* = -T$. Most lie between these extremes, their deviation from self-adjointness measuring how they distort the inner product structure. This interplay between transformation and adjoint will guide our development of orthogonal transformations in the sections ahead.

5.5   Orthogonal Transformations

The most elegant transformations preserve geometric structure. A rotation changes perspective while maintaining shape; a reflection inverts orientation while preserving angles. Such transformations – those respecting the inner product structure we have so carefully built – arise throughout engineering, from rigid body mechanics to quantum measurements to data analysis. Their power lies in a perfect fusion of geometric intuition and algebraic precision.

Definition 5.15 (Orthogonal Transformation). A linear transformation $T : V \to V$ on an inner product space is *orthogonal* if it preserves inner products:

$$\langle Tu, Tv \rangle = \langle u, v \rangle$$

for all vectors $u, v \in V$.   ●

An orthogonal transformation preserves lengths and angles: for all u and v,

$$\|Tv\| = \|v\| \quad : \quad \cos \theta = \frac{\langle Tu, Tv \rangle}{\|Tu\|\|Tv\|} = \frac{\langle u, v \rangle}{\|u\|\|v\|}$$

These geometric constraints have powerful algebraic consequences, connecting our work here to the theory of adjoints developed in the previous section:

Lemma 5.16. *For a linear transformation T on a finite-dimensional inner product space, the following are equivalent:*

1. *T is orthogonal*
2. *$T^*T = TT^* = I$*
3. *$T^* = T^{-1}$*

When we choose orthonormal bases for our space, orthogonal transformations have particularly elegant matrix representations. A square matrix Q is *orthogonal* if

$$Q^TQ = I = QQ^T \tag{5.3}$$

The set of all $n \times n$ orthogonal matrices is denoted $O(n)$. Such matrices inherit remarkable properties that make them particularly valuable in computation:

Lemma 5.17. *An orthogonal matrix Q satisfies:*

1. *Its columns (and rows) form an orthonormal basis*
2. *Its inverse equals its transpose: $Q^{-1} = Q^T$*
3. *It preserves lengths: $\|Qx\| = \|x\|$*
4. *It preserves angles: $(Qx)^T(Qy) = x^Ty$*
5. *Its determinant is ± 1*

Example 5.18 (Rigid Body Motion). The orientation of a rigid body in three-dimensional space is described by an orthogonal transformation. Its

Note: The term "orthogonal" here connotes preservation of all inner products, not merely orthogonality relations. A more accurate (though less traditional) name might be "inner-product-preserving" or "orthonormal."

Foreshadowing: The power of orthogonal matrices in computation will become even clearer when we study the Singular Value Decomposition in Chapter 10, where they provide optimal coordinate transformations for a variety of purposes.

matrix representation R in any orthonormal basis satisfies $R^T R = RR^T = I$, with $\det R = \pm 1$. When $\det R = 1$, the transformation represents a pure rotation; when $\det R = -1$, it includes a reflection.

This geometric structure explains why different physical quantities transform differently under rotation. Position vectors transform by R, while angular momentum vectors transform by $R^{-1} = R^T$, preserving their geometric relationships while respecting the physics of rigid motion. ◇

These mechanical constraints – preservation of distances and angles – arise from the physical principle that rigid bodies maintain their shape under motion. The orthogonality of R encodes this fundamental geometric requirement algebraically.

Example 5.19 (Signal Transforms). The Discrete Fourier Transform (DFT), properly normalized, provides an orthogonal transformation on the space of discrete signals. Its orthogonality manifests as Parseval's identity – the conservation of signal energy between time and frequency domains:

$$\sum_{n=0}^{N-1} |x[n]|^2 = \frac{1}{N} \sum_{k=0}^{N-1} |X[k]|^2$$

where $x[n]$ is the time-domain signal and $X[k]$ its frequency-domain transform. This conservation arises precisely because the DFT basis vectors form an orthonormal set. ◇

Relax... If this makes sense, cool. If not, no worries! *You are having a dream...*

Example 5.20 (Data Analysis). In high-dimensional data analysis, orthogonal transformations provide optimal coordinate changes for revealing structure. Principal Component Analysis (PCA) – which we shall study in detail in Chapter 11 – seeks an orthogonal change of coordinates that aligns the data's axes of maximum variation. The orthogonality ensures that the new coordinates remain uncorrelated, while the inner product preservation guarantees no information is lost in the transformation. ◇

The composition of orthogonal transformations is orthogonal, and the inverse of an orthogonal transformation is orthogonal. This closure under composition and inversion hints at deeper algebraic structure – the orthogonal transformations form a group under composition, though we shall not pursue this abstraction further.

The power of orthogonal transformations lies in their fusion of geometric and algebraic properties. They provide the mathematical framework for rigid motion in mechanics, preserve energy in wave equations and quantum systems, and offer optimal coordinate changes for data analysis. Their ubiquity in engineering practice – from robotics to signal processing to machine learning – reflects a deeper truth: the most useful transformations are often those that preserve fundamental structure.

Caveat: The term "group" here has precise mathematical meaning, describing a set closed under an associative operation, with identity and inverses. Though we won't develop this theory, it explains much about how orthogonal transformations combine.

5.6   The QR Decomposition

Our development of orthogonality through Section 5.3 provides the foundation for another fundamental matrix factorization. The Gram-Schmidt process transforms any basis into an orthogonal one through systematic projection – a process that itself defines a natural decomposition of the original matrix. This factorization, called the QR decomposition, expresses any matrix as a product of an orthogonal matrix and an upper triangular one.

Definition 5.21 (QR Decomposition). The QR *decomposition* of a square matrix $A \in \mathbb{R}^{n \times n}$ expresses it as a product

$$A = QR$$

where Q is orthogonal ($Q^T Q = I$) and R is upper triangular. When A has full rank, this decomposition is unique if we require the diagonal entries of R to be positive.   •

Consider how this decomposition arises from orthogonalizing A's columns. Writing $A = [a_1 \ \cdots \ a_n]$, the Gram-Schmidt process produces orthonormal vectors $\{q_1, \ldots, q_n\}$ where each q_k emerges from a_k by subtracting its projections onto previous vectors. The coefficients of these projections, together with the normalization factors, assemble naturally into an upper triangular matrix R. This construction reveals both why Q must be orthogonal and why R assumes triangular form – each column of A expresses through the q_i only up to its own index.

Example 5.22 (Simple QR Form). Consider the matrix

$$A = \begin{bmatrix} 3 & -4 \\ 4 & 3 \end{bmatrix}$$

Direct computation yields

$$Q = \begin{bmatrix} 0.6 & -0.8 \\ 0.8 & 0.6 \end{bmatrix} \quad \text{and} \quad R = \begin{bmatrix} 5 & 0 \\ 0 & 5 \end{bmatrix}$$

The orthogonal factor Q captures the rotation inherent in A, while the triangular factor R represents scaling. This geometric decomposition – into pure rotation followed by scaling – exemplifies how matrix factorizations reveal underlying structure.   ◇

The QR decomposition provides more than mere factorization – it reveals the matrix's action through natural stages. Like the eigendecomposition studied in Chapter 7 and the singular value decomposition to come in Chapter 10, QR decomposes a transformation into simpler, geometrically meaningful operations:

1. The orthogonal factor Q provides a new orthonormal basis
2. The triangular factor R describes coordinates in this basis
3. Their product QR reconstructs the original transformation

Lemma 5.23 (Existence and Uniqueness). *Every nonsingular matrix A admits a unique QR decomposition with positive diagonal entries in R. When A is singular, the decomposition exists but loses uniqueness in ways determined by the matrix's rank.*

This factorization provides powerful tools for both theoretical analysis and practical computation. Systems of equations yield naturally to solution through QR factors: if $Ax = b$, then $Rx = Q^T b$ reduces to back-substitution. The orthogonal factor Q preserves lengths and angles while R implements a simple triangular transformation. Like a sculptor's fundamental techniques, mastery of the QR decomposition enables increasingly sophisticated applications through careful composition of simple operations.

The power of this decomposition emerges from its fusion of geometric and algebraic perspectives. Though discovered through algebraic orthogonalization, its true power flows from the geometric insight that any linear transformation factors naturally through orthogonal bases. This unity of geometry and algebra – a theme that has guided our development throughout – achieves particular clarity in the QR decomposition, where abstract orthogonality principles transform into practical computational tools.

• ———————————————————————————————————— •

Text Embeddings & Semantic Search

The representation of text as vectors in high-dimensional spaces provides a striking application of inner product geometry. By encoding words or phrases as vectors in $\mathbb{R}^n$ (typically with n very large), we can measure semantic similarity through the geometric tools developed in this chapter. The inner product structure (Definition 5.1) transforms abstract notions of meaning into concrete computational methods, enabling modern language models to process and understand text through geometric operations.

The key insight is that the angle between embedding vectors, measured through their normalized inner product, captures semantic similarity. For embedding vectors $v, w \in \mathbb{R}^n$, their *cosine similarity* is defined as:

$$\cos\theta = \frac{\langle v, w \rangle}{\|v\|\,\|w\|}$$

This measure, ranging from -1 to 1, emerges naturally from the geometric structure we established in Section 5.2. Words or phrases with similar meanings yield embedding vectors with small angles between them, while unrelated concepts become nearly orthogonal in the embedding space.

Algorithm 1: Semantic Search via Cosine Similarity

Data: Query embedding q, Database of embeddings $\{d_1, \ldots, d_N\}$
Result: Indices of k most similar items
Normalize query: $q \leftarrow q/\|q\|$;
for $i \leftarrow 1$ **to** N **do**
   Normalize database vector: $d_i \leftarrow d_i/\|d_i\|$;
   Compute similarity: $s_i \leftarrow \langle q, d_i \rangle$;
end
Return indices of k largest similarities;

Caveat: The effectiveness of embedding-based similarity depends crucially on the quality of the embedding vectors. Modern language models learn these embeddings through extensive training on large text corpora, optimizing the geometric structure to reflect semantic relationships.

Example 5.24 (Word Analogies). A remarkable feature of well-trained embedding spaces is their capture of semantic relationships through vector arithmetic. The classic example

$$\text{king} - \text{man} + \text{woman} \simeq \text{queen}$$

demonstrates how the geometry of the embedding space encodes meaningful relationships: for a semantically suitable text embedding, the vector difference between *king* and *man* embeddings, when added to *woman*, produces a vector close (in cosine similarity) to the embedding of *queen*.   ◇

Example 5.25 (Document Retrieval). Given a corpus of documents, each encoded as a vector through averaging or sophisticated pooling of word embeddings, semantic search reduces to finding nearest neighbors under cosine similarity. This geometric approach captures topical similarity far better than traditional keyword matching, as documents using different but semantically related terms still yield vectors with small angular separation.   ◇

Foreshadowing: In Chapter 10, we shall see how the Singular Value Decomposition enables efficient approximate nearest neighbor search in high-dimensional embedding spaces through dimensionality reduction.

The power of this geometric approach extends beyond simple similarity search. Modern language models construct multiple layers of embedding spaces, each capturing different aspects of linguistic structure. The principles developed in this chapter — from the abstract properties of inner products (Definition 5.1) to the concrete geometry of angles (Lemma 5.16) — provide the mathematical foundation for these sophisticated natural language processing systems.

The success of embedding-based methods in natural language processing exemplifies a broader principle: many complex relationships in data can be captured through careful geometric structure. Whether comparing documents, analyzing molecular structures, or processing social networks, the inner product geometry we have developed offers a powerful framework for measuring and exploiting similarity.

•————————————————————————•

Quantum Measurement & Observable Operators

Quantum mechanics provides perhaps the most elegant application of inner product geometry – not as mere mathematical convenience, but as fundamental physical law. The mathematical structures developed in this chapter – inner products, orthogonality, and adjoints – emerge naturally as the language for describing quantum states and their measurement. For engineering students who have encountered quantum mechanics through physics, this section offers a fresh perspective grounded in the clarity of finite-dimensional linear algebra.

The key insight is that quantum states are simply vectors in an inner product space. A quantum bit or *qubit* – the fundamental unit of quantum information – lives in a two-dimensional complex inner product space $\mathbb{C}^2$. Any state can be written as a linear combination of two orthonormal basis states, traditionally denoted $|0\rangle$ and $|1\rangle$ (physics notation for vectors which we shall embrace briefly):

$$|\psi\rangle = \alpha\,|0\rangle + \beta\,|1\rangle, \quad \text{where } |\alpha|^2 + |\beta|^2 = 1$$

The normalization condition $\|\,|\psi\rangle\,\| = 1$ reflects a fundamental physical principle: probabilities must sum to one.

When we measure a quantum state, we are effectively computing inner products. The probability of measuring the system in state $|\phi\rangle$ when it is in state $|\psi\rangle$ equals the squared magnitude of their inner product:

$$\mathbb{P}(\phi|\psi) = |\langle\phi|\,|\psi\rangle|^2$$

This geometric interpretation transforms quantum measurement from mysterious phenomenon into concrete computation: we simply project our state vector onto various measurement directions and compute probabilities from the resulting inner products.

More general measurements correspond to *observable operators* – linear transformations that are their own adjoints (called *Hermitian* in physics when complex). The eigenvalues of these operators give the possible measurement outcomes, while their eigenvectors provide the corresponding measurement bases. This connection between adjoints and physical measurement provides fundamental constraints on what can be observed in quantum systems.

Example 5.26 (Quantum Bit Measurement). Consider measuring a qubit in state

$$|\psi\rangle = \frac{1}{\sqrt{2}} \begin{pmatrix} 1 \\ 1 \end{pmatrix}$$

in the standard basis. This state assigns equal probability to measuring either 0 or 1:

$$\mathbb{P}(0|\psi) = |\langle 0|\,|\psi\rangle|^2 = \left| \left\langle \begin{pmatrix} 1 \\ 0 \end{pmatrix}, \frac{1}{\sqrt{2}} \begin{pmatrix} 1 \\ 1 \end{pmatrix} \right\rangle \right|^2 = \frac{1}{2}$$

and similarly for $\mathbb{P}(1|\psi)$. The orthonormality of basis states ensures these probabilities sum to one. ◇

A more general measurement corresponds to an observable operator like

$$A = \begin{bmatrix} 1 & 0 \\ 0 & -1 \end{bmatrix}$$

This operator is clearly self-adjoint ($A^* = A$) and has eigenvalues ± 1 with corresponding normalized eigenvectors:

$$v_+ = \begin{pmatrix} 1 \\ 0 \end{pmatrix}, \quad v_- = \begin{pmatrix} 0 \\ 1 \end{pmatrix}$$

When we measure this observable, we will obtain either $+1$ or -1, with probabilities determined by projecting our state onto these eigenvectors.

This framework extends naturally to higher-dimensional systems. A register of n qubits lives in a 2^n-dimensional complex vector space – explaining both the power and fragility of quantum computation. Each additional qubit doubles the dimension of the state space, enabling massive parallelism when manipulating quantum states. Yet this exponential scaling also explains why quantum states are so difficult to simulate classically: representing even 50 qubits requires storing 2^{50} complex numbers.

The principles of Section 5.3 find direct application in quantum error correction. By encoding logical qubits into higher-dimensional spaces using orthogonal subspaces, we can detect and correct certain types of errors through projection operations. The mathematical machinery of inner products and orthogonal complements provides the theoretical foundation for making quantum computation robust against noise.

Example 5.27 (Error Detection). A simple quantum error detecting code encodes a single logical qubit $\alpha \, |0\rangle + \beta \, |1\rangle$ into a two-qubit state:

$$\alpha \, |00\rangle + \beta \, |11\rangle$$

This encoding uses orthogonal states $|00\rangle$ and $|11\rangle$, reserving $|01\rangle$ and $|10\rangle$ as error indicators. When an error occurs, it typically moves the state into the orthogonal complement of the code space – allowing detection through projection measurements. ◇

The study of quantum systems exemplifies how the geometric structures developed in this chapter emerge naturally in physical law. Inner products determine measurement probabilities, adjoints constrain observable operators, and orthogonal subspaces enable error correction. This unity of mathematics and physics provides both practical tools for quantum engineering and deeper insight into the role of geometry in physical law.

For the engineering student, quantum mechanics thus becomes not a collection of mysterious postulates, but a natural application of the linear algebraic concepts developed systematically in this text. The inner product structure first encountered abstractly in Section 5.1 reveals itself as fundamental to the behavior of physical systems at their smallest scales.

Caveat: The restriction to self-adjoint operators for quantum observables emerges from a physical requirement: measurement outcomes must be real numbers. The eigenvalues of a self-adjoint operator are always real, providing exactly this guarantee.

Exercises: Chapter 5

1. Consider the inner product on $\mathcal{P}_2$ given by $\langle f, g \rangle = \int_0^1 f(x)g(x)\,dx$. Find the norm of $p(x) = 1 + 2x + x^2$.

2. Let V be the space of 2×2 matrices with the inner product $\langle A, B \rangle = \operatorname{tr}(A^T B)$. Find the angle between matrices $A = \begin{bmatrix} 1 & 0 \\ 0 & 1 \end{bmatrix}$ and $B = \begin{bmatrix} 1 & -1 \\ 1 & 1 \end{bmatrix}$.

3. Show that $\langle f, g \rangle = f(0)g(0) + f(1)g(1)$ defines an inner product on $\mathcal{P}_1$. Find a vector orthogonal to $p(x) = x$.

4. On $\mathbb{R}^3$, define $\langle u, v \rangle_w = 2u_1 v_1 + 3u_2 v_2 + u_3 v_3$. Prove this is an inner product and find vectors of unit length.

5. For matrices A and B, prove that $\operatorname{tr}(A^T B) = \operatorname{tr}(B^T A)$ and use this to show the Frobenius inner product is symmetric.

6. Let $u = (1, 2, 2)^T$ and $v = (2, 4, -1)^T$. Find an orthonormal basis for $\operatorname{span}\{u, v\}$ using the Gram-Schmidt process, then extend it to an orthonormal basis of $\mathbb{R}^3$. Write the resulting orthogonal matrix Q and verify $Q^T Q = I$.

7. For integers $m, n \geq 0$, prove that the functions $\cos(mx)$ and $\sin(nx)$ are orthogonal on $[-\pi, \pi]$ under the standard L^2 inner product.

8. Let $C([0, 2\pi])$ denote continuous functions on $[0, 2\pi]$ with inner product $\langle f, g \rangle = \int_0^{2\pi} f(x)g(x)\, dx$. Show that $\{1, \cos x, \sin x\}$ forms an orthogonal set but not an orthonormal set. Find the appropriate scaling factors to make it orthonormal.

9. Consider the space SYM_2 of 2×2 symmetric matrices with the Frobenius inner product. Find an orthonormal basis for this space, and verify it has the correct dimension.

10. Let A be the matrix:
$$A = \begin{bmatrix} 1 & 1 \\ 1 & 0 \\ 0 & 1 \end{bmatrix}$$

Find its QR decomposition by applying Gram-Schmidt orthogonalization to its columns. Verify that $Q^T Q = I$ and $A = QR$.

11. Show that if Q is a 2×2 orthogonal matrix with $\det Q > 0$, then Q must be a rotation matrix. If instead $\det Q < 0$, what can you say about it?

12. For points in the plane, define $\langle p, q \rangle_M = p_1 q_1 + p_1 q_2 + p_2 q_1 + 2p_2 q_2$. Write the matrix M such that this inner product equals $p^T M q$. Is this a valid inner product? Justify your answer. If valid, find two orthogonal vectors under this inner product.

13. Define an inner product on $\mathbb{R}^2$ by $\langle u, v \rangle = 4u_1 v_1 + 4u_1 v_2 + 4u_2 v_1 + 5u_2 v_2$. Find the angle between $u = (1, 0)^T$ and $v = (0, 1)^T$ under this inner product.

14. Consider a nonzero vector $v \in \mathbb{R}^3$ and let $\Pi = \Pi_v$ denote orthogonal projection onto $\operatorname{span}\{v\}$. Show that Π is symmetric: $\Pi^T = \Pi$.

15. For vectors $u, v \in \mathbb{R}^n$, prove that the Cauchy-Schwarz inequality becomes an equality if and only if one vector is a scalar multiple of the other.

16. Prove that if u and v are orthogonal vectors in an inner product space, then the Pythagorean theorem holds: $\|u + v\|^2 = \|u\|^2 + \|v\|^2$.

17. An inner product satisfies $\|u + v\|^2 + \|u - v\|^2 = 2(\|u\|^2 + \|v\|^2)$ for all vectors u, v. Prove this using properties of inner products.

18. Let A and B be 2×2 matrices. Using the Frobenius inner product $\langle A, B \rangle = \operatorname{tr}(A^T B)$, show that if A and B are orthogonal as vectors in matrix space, then $\operatorname{tr}(AB^T) = 0$. Is there a geometric interpretation of this orthogonality in terms of how A and B act on vectors in $\mathbb{R}^2$?

19. Let $T : V \to V$ be a linear transformation on an inner product space. Prove that if $\langle Tv, v \rangle = 0$ for all $v \in V$, then $T = 0$.

20. Let $V = \mathbb{R}^2$ with the standard inner product. For a fixed nonzero vector a, define $T : V \to V$ by:

$$Tx = x - 2\frac{\langle x, a \rangle}{\|a\|^2} a$$

Prove that T preserves inner products: $\langle Tx, Ty \rangle = \langle x, y \rangle$ for all $x, y \in V$.

21. For a subspace $W < V$ of an inner product space and vector $v \in V$, prove that:

$$\|v - \Pi_W v\| \leq \|v - w\|$$

for any $w \in W$, with equality if and only if $w = \Pi_W v$. Interpret this result geometrically.

22. The temperature at point (x, y) on a plate is given by $T(x, y) = e^{-x^2 - y^2}$. Using the inner product $\langle f, g \rangle = \int_{-\infty}^{\infty} \int_{-\infty}^{\infty} f(x, y) g(x, y) \, dx \, dy$, find $\|T\|$.

23. The *Gram matrix* G of vectors $\{v_1, \ldots, v_n\}$ has entries $g_{ij} = \langle v_i, v_j \rangle$. Prove that G is: symmetric and has determinant zero if and only if the vectors are linearly dependent.

24. Let V be the space of continuous functions on $[0, 1]$ with inner product $\langle f, g \rangle = \int_0^1 f(x) g(x) \, dx$. Find constants a and b such that $f(x) = a + bx$ is orthogonal to both 1 and x^2.

25. For continuous functions on $[0, 1]$, consider:

$$\langle f, g \rangle = \int_0^1 f(x) g(x) \, dx + f(0) g(0) + f(1) g(1)$$

Prove this defines an inner product and find the norm of $f(x) = x(1 - x)$.

26. On the space $C^1([0, 1])$ of continuously differentiable functions, define:

$$\langle f, g \rangle = \int_0^1 (f(x) g(x) + f'(x) g'(x)) \, dx$$

Prove this defines an inner product and find the angle between $f(x) = x$ and $g(x) = x^2$.

27. Let $\mathcal{P}_3[0, 1]$ denote polynomials of degree at most 3 with inner product $\langle f, g \rangle = \int_0^1 f'(x) g'(x) \, dx$. Prove this defines a valid inner product on the subspace of polynomials with $f(0) = 0$; then, find a polynomial orthogonal to both x and x^2 under this inner product.

Chapter 6
Orthogonal Decomposition & Data

"quadrangular the building rose the heavens squared by a line"

THE MARRIAGE OF GEOMETRY AND ALGEBRA reaches its first culmination in the notion of orthogonality. The inner product structure we have built transforms our understanding of the fundamental spaces and operations from Chapter 3. What was mere algebra – kernels and images, quotients and complements – acquires geometric meaning through perpendicularity. This geometric perspective not only illuminates the theory but guides computation, leading to optimal methods for solving systems and fitting data.

Our task is to reexamine the foundations laid in earlier chapters through the lens of orthogonality. The kernel of a transformation is not merely algebraically invisible but geometrically perpendicular to its coimage. Quotient spaces are not only algebraic identifications but induce orthogonal decompositions. The Fundamental Theorem itself expresses not merely dimensional accounting but geometric splitting of domain and codomain into perpendicular factors.

This geometric reinterpretation bears immediate practical fruit. When exact solutions to linear systems do not exist, orthogonal projections provide best approximations under natural measures of error. Data corrupted by noise finds clean representation through projection onto suitable subspaces. The abstract machinery of inner products descends from theory to practice, offering optimal methods for a host of engineering problems.

The path ahead revisits familiar territory with new eyes. We begin by reinterpreting the four fundamental spaces through orthogonality. This geometric understanding leads naturally to projection operators – the workhorses of decomposition. These tools in hand, we return to the Fundamental Theorem, as emanation of orthogonal decomposition of spaces. Finally, we apply this geometric machinery to the practical problems of least squares approximation, where theory guides us to optimal solutions of overdetermined systems.

This synthesis of geometry and algebra – of inner products and linear transformations – provides the foundation for many modern computational methods. The principles we develop here will guide our study of eigenvalues, singular values, and principal components in chapters ahead.

Foreshadowing: The orthogonal projections studied here form the theoretical foundation for dimensionality reduction in machine learning, where high-dimensional data is projected onto lower-dimensional subspaces that capture essential features.

6.1   Orthogonal Subspaces & Complements

The vector spaces we have studied thus far acquire richer structure through the inner product geometry of Chapter 5. Just as individual vectors may stand perpendicular to one another, entire subspaces can be orthogonal, leading to natural geometric decompositions that both illuminate theory and enable computation.

Definition 6.1 (Orthogonal Complement). Two subspaces U, W of an inner product space V are *orthogonal*, denoted $U \perp W$, if every vector in one is orthogonal to every vector in the other:

$$\langle u, w \rangle = 0 \quad \text{for all } u \in U,\ w \in W$$

The *orthogonal complement* of a subspace $U < V$, denoted $U^\perp$, is the subspace of all vectors orthogonal to U:

$$U^\perp = \{v \in V : \langle v, u \rangle = 0 \text{ for all } u \in U\}$$

That $U^\perp$ is indeed a subspace follows readily from properties of the inner product: the zero vector is certainly orthogonal to all of U, and linear combinations of vectors orthogonal to U remain orthogonal to U.

Lemma 6.2. *For V a finite-dimensional inner product space and any $U < V$:*

1. $(U^\perp)^\perp = U$
2. $\dim U + \dim U^\perp = \dim V$
3. $U \cap U^\perp = \{0\}$
4. $V = U \oplus U^\perp$

Proof. That $U \subseteq (U^\perp)^\perp$ is clear from definitions. For the reverse inclusion, choose any orthonormal basis $\{u_1, \ldots, u_k\}$ for U and extend it to an orthonormal basis of V. The complement of this basis spans $U^\perp$, from which the dimensional equality follows. The remaining properties emerge from this explicit construction. $\square$

The last property is especially significant: every vector in V decomposes uniquely into orthogonal components lying in U and $U^\perp$. We will write $V = U \boxplus U^\perp$ to denote an *orthogonal direct sum* of V into subspaces U and $U^\perp$ that are both complementary and orthogonal. This $\boxplus$-decomposition will prove fundamental to our development of projection operators in the next section.

Example 6.3 (Matrix Subspaces). Consider the space $\mathbb{R}^{n \times n}$ with the Frobenius inner product $\langle A, B \rangle = \mathrm{tr}(A^T B)$ from Example 5.3. The subspace SYM_n of symmetric matrices and the subspace SKEW_n of skew-symmetric matrices are orthogonal complements:

$$\mathbb{R}^{n \times n} = \mathrm{SYM}_n \boxplus \mathrm{SKEW}_n$$

$$\mathrm{SYM}_n = \{A : A^T = A\} \quad \text{and} \quad \mathrm{SKEW}_n = \{A : A^T = -A\}$$

This orthogonal decomposition reveals that any matrix A splits uniquely into symmetric and skew-symmetric parts:

$$A = \frac{A + A^T}{2} + \frac{A - A^T}{2}$$

This decomposition has important applications in mechanics, where symmetric matrices often represent stress and strain.   ◇

The geometric perspective offered by orthogonal complements will transform our understanding of the fundamental subspaces from Chapter 3. What appeared there as purely algebraic constructions – kernels and images, quotients and duals – will acquire new geometric meaning through orthogonality. This reinterpretation through perpendicularity will guide our development of optimal methods for solving systems and approximating data in the sections ahead.

6.2   Projections & Quotients

The concept of projection pervades mathematics and engineering. A shadow cast by sunlight projects three-dimensional objects onto the plane; a surveyor's map projects the curved surface of Earth onto flat paper; a statistician projects high-dimensional data onto informative lower-dimensional summaries. These diverse examples share a common mathematical essence: the approximation of complex objects by simpler ones through systematic dimension reduction.

The inner product structure we have built transforms this intuitive notion into precise mathematics. Given a subspace $W < V$, we seek to approximate arbitrary vectors in V by their "shadows" in W – vectors that minimize the distance to the original while lying entirely in W. This geometric problem leads directly to orthogonal projection, a concept that unifies the theoretical structure of Chapter 3 with the computational methods of modern data analysis.

Definition 6.4 (Orthogonal Projection). Let $W < V$ be a subspace of an inner product space. The *orthogonal projection* onto W is the linear transformation $\Pi_W : V \to V$ satisfying:

1. $\Pi_W v \in W$ for all $v \in V$ (projects onto W)
2. $v - \Pi_W v \perp W$ for all $v \in V$ (projects orthogonally)

Nota bene: The projection Π_W is uniquely determined by these properties. Though other transformations might map vectors into W, only orthogonal projection maintains perpendicularity of the error.

This seemingly abstract definition encodes a powerful optimization principle: $\Pi_W v$ provides the best approximation to v within W under the natural distance measure induced by the inner product.

Lemma 6.5 (Best Approximation). *For any $v \in V$ and $w \in W$:*

$$\|v - \Pi_W v\| \leq \|v - w\|$$

with equality if and only if $w = \Pi_W v$.

Proof. For any $w \in W$, the error vector $v - w$ decomposes into orthogonal components:

$$v - w = (v - \Pi_W v) + (\Pi_W v - w)$$

where the first term is orthogonal to W and the second lies in W. By the Pythagorean theorem:

$$\|v - w\|^2 = \|v - \Pi_W v\|^2 + \|\Pi_W v - w\|^2$$

The right term vanishes precisely when $w = \Pi_W v$. $\square$

When W has an orthonormal basis $\{w_1, \ldots, w_k\}$, the projection takes an especially elegant form:

$$\Pi_W v = \sum_{i=1}^{k} \langle v, w_i \rangle w_i$$

Each coefficient emerges naturally as the inner product with the corresponding basis vector – no system of equations need be solved. This formula reveals projection as a type of spectral decomposition, extracting from v precisely those components aligned with W's basis vectors.

Example 6.6 (Signal Processing). Consider the space $V = C([-\pi, \pi])$ of continuous functions on $[-\pi, \pi]$ with the L^2 inner product of Example 5.9. The subspace W spanned by $\{1, \cos x, \sin x\}$ captures the DC and first harmonic components of signals. The projection Π_W implements a basic low-pass filter, approximating arbitrary signals by their first Fourier components. The error $v - \Pi_W v$ represents higher-frequency content filtered out by the projection. ◇

Projection operators possess several properties that illuminate their geometric and algebraic character:

Lemma 6.7 (Projection Properties). *The orthogonal projection Π_W satisfies:*

1. *Idempotence:* $\Pi_W^2 = \Pi_W$
2. *Self-adjointness:* $\Pi_W^* = \Pi_W$
3. *Complementarity:* $I - \Pi_W = \Pi_{W^\perp}$

where $W^\perp$ denotes the orthogonal complement of W.

These properties reflect the geometric nature of projection – applying it twice has no additional effect; it respects the inner product structure; and it decomposes space into orthogonal pieces. The last property provides particular insight: projection onto W and projection onto $W^\perp$ split any vector into complementary components.

Foreshadowing: This decomposition principle will reach its full power in Chapter 10, where the Singular Value Decomposition provides an optimal sequence of orthogonal projections for approximating data.

This splitting illuminates our earlier study of quotient spaces. When we quotient V by a subspace U, each equivalence class consists of vectors differing by elements of U. Through the lens of orthogonality, we can represent each class by its projection onto $U^\perp$ – the unique member minimizing distance to the origin. The quotient space V/U becomes naturally isomorphic to $U^\perp$, with the projection $\Pi_{U^\perp}$ providing an explicit isomorphism.

The relationship between projections and quotients can be visualized through a commutative diagram. The natural quotient map π (dashed) makes the diagram commute: both paths from V to W yield equivalent results.

$$\begin{array}{ccc} V & \overset{\pi}{\dashrightarrow} & V/U \\ \Big\downarrow{\scriptstyle \Pi_{U^\perp}} & & \Big\downarrow{\scriptstyle \varphi} \\ U^\perp & \overset{\psi}{\longrightarrow} & W \end{array}$$

Nota bene: The dashed arrow represents the natural quotient map, while solid arrows show explicit linear transformations.

Here V/U denotes the quotient space, $U^\perp$ is the orthogonal complement, and W is any space isomorphic to both (often taken to be the image of a linear transformation). The diagram commutes in that both paths from V to W yield the same result – whether we first project onto $U^\perp$ or pass to the quotient space V/U. This geometric perspective illuminates why quotient spaces and orthogonal complements provide equivalent ways of understanding the effective domain of a linear transformation.

Example 6.8 (Data Centering). Consider a collection of vectors $\{x_1, \ldots, x_n\}$ in $\mathbb{R}^d$. The subspace U spanned by $\mathbf{1} = (1, \ldots, 1)^T$ represents uniform translations. Projecting onto $U^\perp$ centers the data by subtracting means – a fundamental preprocessing step in data analysis. The quotient $\mathbb{R}^d / U$ captures the intrinsic shape of the data cloud independent of its absolute position. ◇

The power of orthogonal projection extends far beyond these elementary examples. When exact solutions to linear systems do not exist, projection onto appropriate subspaces yields optimal approximations. When data contains noise, projection onto signal subspaces enables filtering and compression. When complex systems require simplified models, projection onto lower-dimensional spaces balances accuracy and complexity. These applications, and many more, spring from the simple geometric principle encoded in Definition 6.4.

6.3   The Fundamental Theorem Redux

The Fundamental Theorem of Linear Algebra, when viewed in finite-dimensional inner product spaces, reveals deeper structure through the lens of orthogonality. What first appeared as a collection of dimensional relationships now emerges in its true form: a statement about the geometric splitting of spaces. The four fundamental subspaces – kernel, image, cokernel, and coimage – relate not merely through counting of dimensions but through perpendicularity. This geometric understanding, though restricted to the finite-dimensional setting, transforms our perspective on linear transformations, providing both theoretical insight and practical methods for computation.

Recall, we write $A \boxplus B$ to denote an orthogonal direct sum: subspaces A and B that are both complementary $(V = A \oplus B)$ and orthogonal $(A \perp B)$.

Theorem 6.9 (Fundamental Theorem of Linear Algebra (Geometric Form)). *Any linear transformation $T : V \to W$ between finite-dimensional inner product spaces induces orthogonal decompositions of both domain and codomain:*

$$V = \ker T \boxplus (\ker T)^\perp \qquad and \qquad W = \operatorname{im} T \boxplus (\operatorname{im} T)^\perp \qquad (6.1)$$

These decompositions are connected by the following:

1. *The restriction of T to $(\ker T)^\perp$ gives an isomorphism $(\ker T)^\perp \cong \operatorname{im} T$*
2. *The coimage $V / \ker T$ is naturally isomorphic to $\operatorname{im} T$*
3. *The cokernel $W / \operatorname{im} T$ is naturally isomorphic to $(\operatorname{im} T)^\perp$*
4. *The orthogonal projections $\Pi_{(\ker T)^\perp}$ and $\Pi_{\operatorname{im} T}$ commute with T*

Nota bene: The term *naturally isomorphic* here means that the isomorphisms arise from the geometric structure itself.

These decompositions illuminate the four fundamental subspaces through geometry rather than algebra. The kernel represents vectors invisible to T; its orthogonal complement captures the effective inputs. The image contains all possible outputs; its orthogonal complement measures the transformation's deficiency. Each complementary pair provides a complete view of how T acts on its domain and codomain.

Example 6.10 (Matrix Transformations). For a matrix $A \in \mathbb{R}^{m \times n}$, these decompositions have immediate computational significance with respect to $\text{ROW}(A)$ and $\text{COL}(A)$, the row and column spaces respectively.

1. $\mathbb{R}^n = \ker(A) \boxplus \text{ROW}(A)^T$ splits input space
2. $\mathbb{R}^m = \text{COL}(A) \boxplus \ker(A^T)$ splits output space
3. The projections $\Pi_{\text{ROW}(A)^T}$ and $\Pi_{\text{COL}(A)}$ provide optimal approximate solutions when exact solutions do not exist

$\diamond$

Foreshadowing: The orthogonal complements appearing here foreshadow the notion of orthogonal matrices in Chapter 7, where entire transformations preserve perpendicularity.

These geometric splittings yield the algebraic relationships of Chapter 3 as corollaries:

Corollary 6.11 (Rank-Nullity Redux). *For a linear transformation $T : V \to W$ between finite-dimensional inner product spaces:*

1. $\dim V = \dim \ker T + \dim \textit{im } T$
2. $\dim W = \dim \textit{im } T + \dim \textit{coker } T$
3. $\dim \textit{im } T = \dim(\ker T)^{\perp}$
4. *The rank equals both* $\dim \textit{im } T$ *and* $\dim \textit{coim } T$

The proof follows from the orthogonal decompositions, with the dimensions of complementary subspaces summing to the dimension of the whole space. What before seemed like mysterious algebraic coincidences now emerge as natural consequences of geometric splitting.

This geometric perspective guides computation. When solving $Tv = w$:

1. Project w onto im T to test solvability
2. If solvable, find a particular solution in $(\ker T)^{\perp}$
3. If unsolvable, project onto im T for best approximation

The Fundamental Theorem thus reveals itself as more than mere algebra – it expresses the fundamental geometric structure of linear transformations. This unity of geometry and algebra provides both theoretical insight and practical methods, a theme that will deepen as we proceed to least squares problems and their applications.

6.4   The Pseudoinverse

The Fundamental Theorem of Linear Algebra revealed how any linear transformation induces four fundamental subspaces, connected through orthogonal decomposition of domain and codomain. This elegant structure suggests a natural question: can we define a reverse transformation that somehow undoes the action of our original map while respecting these geometric relationships? The answer lies in the *pseudoinverse* – a generalization of matrix inversion that provides the optimal approximate inverse even for singular or rectangular matrices.

Definition 6.12 (Pseudoinverse). For a linear transformation $T : V \to W$ between finite-dimensional inner product spaces, the *pseudoinverse* (or *Moore-Penrose inverse*) $T^\dagger : W \to V$ is the unique linear transformation satisfying all of the following conditions:

1. $TT^\dagger T = T$ (First consistency condition)
2. $T^\dagger TT^\dagger = T^\dagger$ (Second consistency condition)
3. $(TT^\dagger)^* = TT^\dagger$ (First adjoint condition)
4. $(T^\dagger T)^* = T^\dagger T$ (Second adjoint condition)

where $*$ denotes the adjoint operation with respect to the inner products on V and W.   •

This abstract definition, while complete, may obscure the profound geometric meaning of the pseudoinverse. Through the lens of the Fundamental Theorem, the pseudoinverse emerges naturally from orthogonal projections onto the four fundamental subspaces:

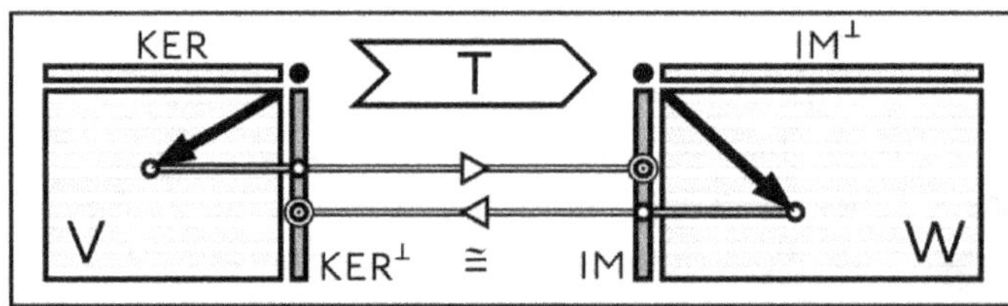

The pseudoinverse $T^\dagger$ as the reverse of T through the fundamental subspaces, with T first projecting onto $(\ker T)^\perp$, then isomorphically mapping to im T; while $T^\dagger$ first projects onto im T, then isomorphically maps to $(\ker T)^\perp$.

This symmetry reveals the pseudoinverse not as an ad hoc construction, but as the natural reverse map that respects the orthogonal structure imposed by our inner products.

For matrices, the pseudoinverse takes a particularly elegant form. When A has full column rank, its pseudoinverse is:

$$A^\dagger = (A^T A)^{-1} A^T$$

This formula connects directly to the orthogonal projections studied earlier in this chapter: $A^T A$ is invertible precisely because A has full column rank, making $\ker(A) = \{0\}$. When A has full row rank, the pseudoinverse becomes:

$$A^\dagger = A^T(AA^T)^{-1}$$

These formulas illustrate how the pseudoinverse generalizes the concept of matrix inversion to rectangular matrices, providing the best possible approximate inverse when an exact inverse doesn't exist.

Example 6.13 (Orthogonal Projection). Consider the orthogonal projection $\Pi_U : V \to V$ onto a subspace $U < V$. Its fundamental spaces are:

- $\ker(\Pi_U) = U^\perp$
- $\mathrm{im}(\Pi_U) = U$

The pseudoinverse $\Pi_U^\dagger$ equals Π_U itself, as projection already satisfies all four pseudoinverse conditions. Indeed, projection is its own pseudoinverse precisely because it is already idempotent ($\Pi_U^2 = \Pi_U$) and self-adjoint ($\Pi_U^* = \Pi_U$). ◇

Example 6.14 (Full Column Rank Matrix). Consider the matrix

$$A = \begin{bmatrix} 1 & 0 \\ 0 & 1 \\ 1 & 0 \end{bmatrix}$$

which has full column rank. Its pseudoinverse is given by:

$$A^\dagger = (A^T A)^{-1} A^T = \begin{bmatrix} 2 & 0 & 1 \\ 0 & 1 & 0 \end{bmatrix}^{-1} \begin{bmatrix} 1 & 0 & 1 \\ 0 & 1 & 0 \end{bmatrix} = \begin{bmatrix} 1/2 & 0 & 1/2 \\ 0 & 1 & 0 \end{bmatrix}$$

This pseudoinverse properly balances the redundant information in the first and third rows of A. ◇

Theorem 6.15 (Pseudoinverse Properties). *The pseudoinverse $T^\dagger$ satisfies:*

1. *$T^\dagger$ maps $\mathrm{im}\,T$ isomorphically to $(\ker T)^\perp$*
2. *$T^\dagger$ maps $(\mathrm{im}\,T)^\perp$ to $\ker T$*
3. *$TT^\dagger$ is the orthogonal projection onto $\mathrm{im}\,T$*
4. *$T^\dagger T$ is the orthogonal projection onto $(\ker T)^\perp$*
5. *If T is invertible, then $T^\dagger = T^{-1}$*
6. *$(T^\dagger)^\dagger = T$*
7. *$(T^*)^\dagger = (T^\dagger)^*$*

In the general case where A is neither full row nor full column rank, the construction of the pseudoinverse becomes more involved. One approach uses the orthogonal projections onto the fundamental subspaces as illustrated in Figure 6.1. First, we project onto the image space $\mathrm{im}(A)$

using the projection operator $P_{\text{im}(A)}$. Then, we apply the isomorphism between $\text{im}(A)$ and $(\ker A)^{\perp}$, followed by the inclusion map back into the domain.

6.5   Least Squares Approximation

The pseudoinverse transforms abstract decompositions into practical computational tools. For systems where exact solutions fail to exist, it navigates the fundamental subspaces to provide optimal approximations. This optimality – finding the best possible approximate solution under natural measures of error – lies at the heart of least squares approximation. The geometric structure revealed through the Fundamental Theorem provides not just theoretical understanding but practical methods for data fitting and analysis.

Consider the system $Ax = b$ where $A : \mathbb{R}^n \to \mathbb{R}^m$ with $m > n$. Such systems typically arise when fitting models to data: each row represents an observation, each column a parameter to be determined. Though b rarely lies in $\text{im}\, A$, the orthogonal decomposition of the codomain $\mathbb{R}^m$ guides us to the optimal approximation:

$$\mathbb{R}^m = \text{im}\, A \boxplus (\text{im}\, A)^{\perp}$$

The vector b thus splits uniquely as $b = b_1 + b_2$ where $b_1 \in \text{im}\, A$ and $b_2 \in (\text{im}\, A)^{\perp}$. Since b_1 lies in $\text{im}\, A$, there exists $\hat{x}$ such that $A\hat{x} = b_1$. This vector $\hat{x}$ provides precisely the least squares solution we seek, as the following theorem confirms:

Theorem 6.16 (Least Squares Solution). *For a full-rank matrix $A \in \mathbb{R}^{m \times n}$ with $m > n$, the system $Ax = b$ has unique least squares solution:*

$$\hat{x} = A^{\dagger}b = (A^T A)^{-1} A^T b$$

This solution minimizes $\|b - Ax\|$ over all $x \in \mathbb{R}^n$.

Proof. The error vector $b - Ax$ must be orthogonal to $\text{im}\, A$ at the minimum, else we could reduce its length through projection. This orthogonality condition means:

$$\langle b - A\hat{x}, Av \rangle = 0 \quad \text{for all } v \in \mathbb{R}^n$$

Therefore $A^T(b - A\hat{x}) = 0$, yielding the *normal equations*:

$$A^T A\hat{x} = A^T b$$

When A has full rank, $A^T A$ is positive definite hence invertible, giving the stated solution. This solution equals $A^{\dagger}b$ by the definition of the pseudoinverse for full-column rank matrices. The pseudoinverse $A^\dagger = (A^T A)^{-1} A^T$ transforms our abstract decomposition into concrete computational methods for finding optimal approximations.

To confirm this minimizes the error, note that for any x:

$$\|b - Ax\|^2 = \|b - A\hat{x} + A\hat{x} - Ax\|^2 = \|b - A\hat{x}\|^2 + \|A\hat{x} - Ax\|^2$$

since $(b - A\hat{x}) \perp \operatorname{im} A$ by construction. The second term vanishes precisely when $x = \hat{x}$, confirming optimality. $\qquad\square$

Nota bene: The term "normal equations" reflects geometric normality – the error vector stands perpendicular to the solution space.

The normal equations emerge naturally from projecting b onto $\operatorname{im} A$. Indeed, the matrix product $A(A^T A)^{-1} A^T$ implements precisely this orthogonal projection. Through the pseudoinverse, we connect the geometric structure of the Fundamental Theorem directly to computational methods for data fitting. This connection transforms abstract subspaces into practical tools for approximation.

Example 6.17 (Linear Regression). Consider fitting a line $y = mx + b$ to points $(x_1, y_1), \ldots, (x_n, y_n)$. This leads to the overdetermined system:

$$\begin{bmatrix} x_1 & 1 \\ x_2 & 1 \\ \vdots & \vdots \\ x_n & 1 \end{bmatrix} \begin{pmatrix} m \\ b \end{pmatrix} = \begin{pmatrix} y_1 \\ y_2 \\ \vdots \\ y_n \end{pmatrix}$$

The least squares solution minimizes the sum of squared vertical distances from points to the line – a criterion that emerges naturally from the Euclidean inner product structure.

For specific values $(x_1, y_1) = (1, 2)$, $(x_2, y_2) = (2, 3)$, and $(x_3, y_3) = (3, 5)$, we form:

$$A = \begin{bmatrix} 1 & 1 \\ 2 & 1 \\ 3 & 1 \end{bmatrix} \quad \text{and} \quad b = \begin{pmatrix} 2 \\ 3 \\ 5 \end{pmatrix}$$

Computing $A^T A$ and $A^T b$:

$$A^T A = \begin{bmatrix} 14 & 6 \\ 6 & 3 \end{bmatrix} \quad \text{and} \quad A^T b = \begin{pmatrix} 23 \\ 10 \end{pmatrix}$$

The normal equations yield:

$$\begin{bmatrix} 14 & 6 \\ 6 & 3 \end{bmatrix} \begin{pmatrix} m \\ b \end{pmatrix} = \begin{pmatrix} 23 \\ 10 \end{pmatrix}$$

Solving, we find $m = 1.5$ and $b = 0.5$, giving $y = 1.5x + 0.5$ as our best-fit line. $\qquad\diamond$

The least squares method extends naturally beyond simple curve fitting. When the columns of A represent basis functions, we obtain general linear models:

$$f(x) = c_1\phi_1(x) + c_2\phi_2(x) + \cdots + c_n\phi_n(x)$$

The coefficients c_i emerge from the least squares solution, providing optimal approximation in the chosen basis. Different choices of basis functions ϕ_i yield different approximation schemes:

- Polynomials for smooth functions
- Trigonometric functions for periodic data
- Wavelets for localized features
- Splines for piecewise smooth approximation

Example 6.18 (Polynomial Fitting). To fit a quadratic $f(x) = ax^2 + bx + c$ to data points (x_i, y_i), we form the Vandermonde matrix:

$$A = \begin{bmatrix} x_1^2 & x_1 & 1 \\ x_2^2 & x_2 & 1 \\ \vdots & \vdots & \vdots \\ x_n^2 & x_n & 1 \end{bmatrix}$$

The least squares solution $c = (a, b, c)^T$ minimizes $\sum_{i=1}^{n}(y_i - f(x_i))^2$. Through the pseudoinverse $A^\dagger$, this solution navigates the fundamental subspaces to provide the optimal approximation. $\diamond$

This geometric understanding transforms least squares from a computational method into a theoretical principle. When exact solutions do not exist, orthogonal projection onto the available solution space provides the best possible approximation under natural measures of error. Through the pseudoinverse, the abstract subspaces of the Fundamental Theorem – the image $\operatorname{im} A$ and its orthogonal complement $(\operatorname{im} A)^\perp$ – translate directly into optimal approximation methods.

The full power of this approach emerges through the explicit connection to the four fundamental subspaces. The least squares solution lives in $(\ker A)^\perp$, making it the unique minimum-norm solution. The residual $b - A\hat{x}$ lies in $(\operatorname{im} A)^\perp$, giving it the geometric interpretation as the component of b that cannot be represented in the model space. This orthogonal decomposition:

$$b = A\hat{x} + (b - A\hat{x})$$

expresses precisely the projection onto the fundamental subspaces guaranteed by the Fundamental Theorem. The pseudoinverse $A^\dagger$ transforms this abstract decomposition into concrete computational methods for finding optimal approximations.

6.6   Regularized Least Squares

Real data harbors noise. The elegant framework of least squares approximation, though mathematically complete, can prove fragile when confronted with measurement errors and uncertainty. Consider fitting a polynomial to noisy samples – increasing the degree improves the fit to our data points but may produce wild oscillations between them. This tension between fidelity to measurements and smoothness of solutions suggests a modification to our geometric framework, one that tames such instabilities while preserving the essential character of orthogonal projection.

The source of trouble lies in our unconstrained pursuit of minimal error. Given noisy measurements, the least squares solution may achieve a deceptively small residual by contorting itself to match the noise rather than the underlying signal. We require some means of favoring simpler, more stable solutions – a preference that we can encode through geometry.

Example 6.19 (Polynomial Overfitting). Consider fitting polynomials of increasing degree to samples of $f(x) = \cos(2\pi x)$ on $[0, 1]$ with small random errors. The least squares solution tracks the data points with increasing precision as degree grows, but at the cost of violent oscillations between samples. Though each fit minimizes squared error, higher-degree solutions appear increasingly unstable.

$\diamond$

Ridge regression provides an elegant solution through a simple modification of our least squares framework. Rather than minimizing only the error $\|Ax - b\|^2$, we add a term penalizing large coefficients:

$$\min_{x} \left(\|Ax - b\|^2 + \lambda \|x\|^2 \right)$$

where $\lambda > 0$ controls the strength of regularization. This augmented objective retains the geometric character of our previous development – it measures not just distance to the data but also distance from the origin in the solution space.

From a theoretical perspective, ridge regression replaces the pseudoinverse $A^\dagger$ with a regularized version $(A^T A + \lambda I)^{-1} A^T$, which sacrifices exact optimality for improved stability and conditioning. This modified

Foreshadowing: The Singular Value Decomposition in Chapter 10 will provide an even more powerful framework for analyzing and solving least squares problems, revealing the full geometric structure of approximate solutions.

Foreshadowing: This balance between fitting data and maintaining simplicity appears throughout mathematics and engineering. We shall encounter it again when studying data analysis in later chapters.

pseudoinverse balances the minimum-norm solution against regularization constraints, effectively shrinking coefficients toward zero while still respecting the underlying orthogonal structure.

The geometric interpretation proves illuminating. Pure least squares projects b onto the column space of A. Ridge regression shifts this projection, pulling solutions toward the origin through an additional orthogonality constraint. The parameter λ controls this shift: larger values favor smaller coefficients at the cost of larger residuals.

Lemma 6.20 (Ridge Solution). *The minimizer of the ridge regression objective satisfies the modified normal equations:*

$$(A^T A + \lambda I)x = A^T b$$

This solution has smaller coefficients than the pure least squares solution but generally larger residual error.

Proof. The objective function $f(x) = \|Ax - b\|^2 + \lambda \|x\|^2$ is minimized where its derivative vanishes. Expanding using properties of inner products:

$$\begin{aligned} f(x) &= \langle Ax - b, Ax - b \rangle + \lambda \langle x, x \rangle \\ &= \langle Ax, Ax \rangle - 2\langle Ax, b \rangle + \|b\|^2 + \lambda \|x\|^2 \end{aligned}$$

Setting the derivative with respect to x to zero:

$$2A^T Ax - 2A^T b + 2\lambda x = 0$$

Rearranging yields the modified normal equations:

$$(A^T A + \lambda I)x = A^T b$$

This regularized system always has a unique solution, even when $A^T A$ is singular, as the added term λI ensures positive definiteness. The solution implements a form of biased pseudoinverse:

$$x = (A^T A + \lambda I)^{-1} A^T b$$

which trades unbiasedness for reduced variance in parameter estimates. The parameter λ controls this bias-variance tradeoff, with larger values producing smaller coefficients but potentially larger residual error. $\qquad\square$

Example 6.21 (Signal Smoothing). Consider smoothing a noisy time series through local polynomial fitting. Pure least squares produces fits that track noise too closely. Ridge regression, by penalizing large coefficients, yields smoother approximations that better capture underlying trends while suppressing high-frequency noise. The parameter λ provides direct control over this smoothing effect. $\qquad\diamond$

The choice of regularization parameter λ embodies the fundamental tradeoff between fitting data and maintaining stability. Small values yield solutions close to pure least squares; large values force solutions toward zero. No universal choice exists – the appropriate balance depends on noise levels, problem structure, and ultimate purpose.

Example: In polynomial fitting, larger λ values increasingly suppress higher-degree terms, effectively limiting the complexity of the fitted function regardless of formal degree.

This modification of least squares – this regularization through additional geometric constraints – exemplifies a broader principle in computational mathematics. When theoretical elegance meets practical complexity, we often find success not by abandoning our framework but by carefully augmenting it. The geometric intuition that guided our development of least squares proves robust enough to accommodate these practical concerns while maintaining its essential character.

Signal Processing: From Data to Information

The passage from raw measurements to meaningful information defines modern signal processing. Every sensor — whether measuring temperature, pressure, acceleration, or electromagnetic waves — provides data whose natural representation lies in a quotient space modulo measurement noise. The mathematical framework we have built in this chapter, particularly orthogonal projection, pseudoinverse computation, and their regularized variants, provides the tools to extract signal from noise through systematic decomposition of spaces.

Consider a temperature sensor sampling at regular intervals. We represent its measurements as a vector $y \in \mathbb{R}^n$ whose components combine true temperature with random fluctuations:

$$y = s + n$$

where s represents the underlying signal and n the noise. Our task is to recover s from y — a challenge that leads naturally to the pseudoinverse operations studied in Section 6.4.

The key insight is that real temperature variations typically occur more slowly than random measurement noise. We can express this mathematically by assuming s lies near a subspace spanned by low-degree polynomials. If we represent our basis as columns of a matrix A, then the optimal recovery becomes:

$$\hat{s} = A(A^T A)^{-1} A^T y = A A^\dagger y$$

This is precisely the orthogonal projection through the pseudoinverse, selecting the optimal point in our model subspace.

For any window of time points $[t_k - w, t_k + w]$, we consider the subspace $\mathcal{P}_d$ of polynomials of degree at most d. The basis polynomials $\{1, t, t^2, \ldots, t^d\}$ provide directions for decomposing our signal, though for numerical stability we often use orthogonal polynomials as discussed in Section 5.3. Each local smoothing operation becomes an application of the pseudoinverse, with ridge regularization providing the stability studied in Section 6.6.

Example 6.22 (Temperature Monitoring). Consider hourly temperature measurements from a chemical reactor represented as $y \in \mathbb{R}^{24}$. Raw sensor data shows rapid fluctuations from measurement noise superimposed on the true temperature trends. The space of quadratic polynomials $\mathcal{P}_2$ provides a natural subspace for local approximation — its three-dimensional structure captures constant levels, linear trends, and gentle curvature while excluding higher-frequency noise.

For our design matrix A whose columns represent basis polynomials evaluated at our sampling points, the pseudoinverse $A^\dagger = (A^T A)^{-1} A^T$ gives us exactly the optimal recovery operator. When the problem becomes ill-conditioned due to closely spaced samples, the regularized pseudoinverse $(A^T A + \lambda I)^{-1} A^T$ stabilizes our solution while maintaining near-optimality. The regularization parameter λ controls the balance between fitting fidelity and coefficient stability, precisely as analyzed in Section 6.6.   ◇

The quotient space perspective proves particularly illuminating. Two temperature signals that differ only by high-frequency noise belong to the same equivalence class in an appropriate quotient space. The pseudoinverse operators we have developed provide a systematic way to select canonical representatives from these equivalence classes — those representatives minimizing both approximation error and coefficient magnitude.

Example 6.23 (Electrocardiogram Processing). An ECG signal represented as $y \in \mathbb{R}^n$ contains both high-frequency noise and sharp features (the QRS complex) that must be preserved. Simple pseudoinverse projection would either retain too much noise or blur important peaks. The solution emerges from adaptive regularization — by varying the regularization parameter λ based on local signal properties, we can tune the pseudoinverse to respect the local structure.

Specifically, in regions with sharp transitions, we reduce regularization to preserve detail, while in flatter regions we increase regularization to suppress noise. This adaptive approach demonstrates how the pseudoinverse framework from Section 6.4 guides practical algorithm design beyond simple least squares. ◇

The framework extends naturally to multidimensional signals through the tensor product constructions implicit in our treatment of matrix spaces. Consider an array of pressure sensors monitoring structural loads on an aircraft wing. Their readings form a matrix $Y \in \mathbb{R}^{m \times n}$, but we expect the true pressure field to vary smoothly across the wing's surface. Two-dimensional polynomial fitting becomes a pseudoinverse problem with a structured design matrix, with regularization ensuring stability just as in the one-dimensional case.

The fundamental principle remains: meaningful information often lives in a lower-dimensional subspace than raw measurements. By carefully choosing these subspaces and using the pseudoinverse and regularized projections developed in this chapter, we separate signal from noise in a mathematically principled way. The geometric intuition we have built — orthogonality, pseudoinverses, and regularized projection — provides the foundation for this essential task of modern engineering.

Image Processing: Pixels & Polynomials

Digital images present a natural domain for applying the orthogonal decomposition and pseudoinverse methods developed in this chapter. Each grayscale image is represented by a matrix $F \in \mathbb{R}^{m \times n}$ whose entries combine meaningful content with measurement noise. The mathematical framework we have built — particularly the pseudoinverse operations of Section 6.4 and regularized least squares of Section 6.6 — provides systematic tools for image enhancement through optimal approximation.

Consider the fundamental problem of image smoothing. The key insight parallels our development of the pseudoinverse: most natural images are locally well-approximated by elements of low-dimensional subspaces. Just as Section 6.4 showed how the pseudoinverse provides optimal projection onto compatible subspaces, we can enhance images through careful application of local pseudoinverse operations.

Example 6.24 (Local Surface Fitting). At each pixel location (i, j), a neighborhood of intensities defines a vector that we fit using the pseudoinverse of an appropriate design matrix. The basis polynomials

$$\{b_{k\ell}(x, y) = (x - i)^k (y - j)^\ell : k + \ell \leq d\}$$

form the columns of our design matrix A. Computing $A^\dagger = (A^T A)^{-1} A^T$ gives us the optimal fitting operator for this polynomial space. As in Section 6.6, the regularized pseudoinverse $(A^T A + \lambda I)^{-1} A^T$ prevents oscillations near edges where the polynomial model becomes strained.   ◇

The geometric interpretation proves particularly illuminating. Each local fit implements the pseudoinverse operators developed in Section 6.4, mapping patches of image data onto carefully chosen polynomial subspaces. The regularization parameter λ shapes this process, pulling solutions toward simpler polynomial coefficients just as ridge regression stabilized the general least squares problems analyzed in Section 6.6.

Example 6.25 (Medical Imaging). Consider X-ray images used in medical diagnosis, represented as matrices $F \in \mathbb{R}^{m \times n}$ containing quantum noise from photon counting statistics. The space of diagnostic features forms a natural quotient modulo this noise structure. The pseudoinverse computation, regularized appropriately to preserve clinically relevant details, provides a systematic way to select canonical representatives from these equivalence classes. The framework of Section 6.4 transforms parameter selection from art to principled mathematics.   ◇

The pseudoinverse perspective introduced in Section 6.4 provides particular insight for image inpainting — reconstructing missing or corrupted pixels. When some pixels are missing, the design matrix A contains rows corresponding only to the observed pixels. The pseudoinverse $A^\dagger$ provides the optimal reconstruction that fits observed data while minimizing coefficient complexity. Since the pseudoinverse minimizes both the residual and the solution norm, it naturally balances fidelity to surrounding data against solution simplicity.

Color images introduce additional structure through their multiple channels. An RGB image comprises three matrices $F_R, F_G, F_B \in \mathbb{R}^{m \times n}$ that benefit from joint analysis. The pseudoinverse methods developed in this chapter extend naturally:

- Channel coupling through appropriate block-structured design matrices
- Joint regularization preserving color consistency
- Pseudoinverse computation respecting inter-channel relationships

Example 6.26 (Satellite Imaging). Satellite imagery provides a striking example of multi-channel data, with each spectral band yielding a matrix $F_k \in \mathbb{R}^{m \times n}$. The meaningful information often lies in a lower-dimensional subspace than the raw measurements. A properly structured pseudoinverse computation, implementing the framework developed in Section 6.4, separates essential features from sensor noise while respecting the coupling between spectral bands.

This multi-channel regularization represents a natural extension of the basic pseudoinverse operations to handle the structured redundancy across different spectral measurements. Just as the pseudoinverse finds the minimum-norm solution to an underdetermined system, these multi-spectral methods find the solution with minimum cross-channel variation.   ◇

• ———————————————————————————————————— •

Support Vector Machines: Geometry of Optimal Separation

The principles of orthogonal decomposition, pseudoinverse computation, and optimal approximation illuminate a fundamental challenge in engineering: how can we best separate data into distinct classes? Support Vector Machines [SVMs] approach this question through the geometry of separating hyperplanes, transforming classification into a problem deeply connected to the pseudoinverse operators and optimization methods developed in this chapter.

Consider points $\{x_i\}_{i=1}^{n}$ in $\mathbb{R}^d$ labeled as either positive ($y_i = +1$) or negative ($y_i = -1$). Just as we sought optimal subspaces for approximating data, we now seek an optimal hyperplane for separating classes. This geometric viewpoint transforms classification from abstract pattern matching into concrete optimization of distances and projections.

A separating hyperplane has form $\{x : w^T x + b = 0\}$ where w provides the normal direction. This geometric object naturally decomposes space into positive and negative halfspaces through the sign of $w^T x + b$. The distance from any point x to this hyperplane emerges through orthogonal projection:

$$d(x) = \frac{|w^T x + b|}{\|w\|}$$

The geometric insight of SVMs lies in maximizing the *margin* — the minimum distance between the hyperplane and any training point. After rescaling w and b to ensure this minimum distance equals $1/\|w\|$, we seek to minimize $\|w\|^2$ subject to correct classification of all points. The resulting optimization mirrors the trade-offs encountered in pseudoinverse computation: we balance the robustness gained from a large margin (analogous to a small solution norm) against the constraints of fitting our training data (analogous to residual minimization).

When data proves inseparable by any single hyperplane, we introduce slack variables ξ_i measuring the degree of misclassification. The optimization becomes:

$$\min_{w,b,\xi} \frac{1}{2}\|w\|^2 + C\sum_{i=1}^{n}\xi_i \quad \text{subject to} \quad y_i(w^T x_i + b) \geq 1 - \xi_i$$

The parameter C controls our balance between margin size and classification errors — exactly paralleling how regularization balanced fit against stability in the pseudoinverse computations of Section 6.4.

Consider how this framework guides medical image classification. Each scan becomes a point in high-dimensional space where coordinates represent intensities, textures, and shapes extracted from the image. The SVM finds an optimal separating hyperplane through principles similar to pseudoinverse computation. The points exactly satisfying the margin constraints — the support vectors — identify the most diagnostically challenging cases. Their special role emerges naturally from the geometry: these boundary cases completely determine the optimal hyperplane, while other points could be moved within their halfspace without affecting the solution.

Nota bene: The support vectors that determine the margin reflect a deeper sparsity principle similar to how the pseudoinverse provides minimum-norm solutions. Both approaches identify the essential components that dictate optimal behavior.

This geometric perspective transforms our approach to structural health monitoring. Vibration sensors on bridges and buildings generate streams of acceleration data requiring rapid classification as either normal or potentially dangerous. Features extracted from these signals become coordinates in a high-dimensional space where the SVM constructs its separating hyperplane. The margin provides crucial tolerance against sensor noise and environmental variation — we need decisions that remain reliable despite uncertainty in our measurements. Points near the margin identify borderline structural states warranting particular attention, while the slack variables allow for occasional constraint violations without compromising overall robustness.

The solution reflects the Fundamental Theorem's decomposition of space as elaborated through the pseudoinverse in Section 6.4. Once we have found our optimal hyperplane, its normal vector w naturally splits $\mathbb{R}^d$ into the direct sum of a line (parallel to w) and a hyperplane (the space of directions parallel to the decision boundary). This geometric decomposition guides both our understanding and our computations: distances to the hyperplane measure confidence in our predictions, while the margin quantifies robustness against input perturbations.

Exercises: Chapter 6

1. For vectors $v_1 = (1,1,1)^T$ and $v_2 = (1,2,-1)^T$ in $\mathbb{R}^3$: (a) Find a vector w that is orthogonal to both v_1 and v_2 (b) Verify that $\{v_1, v_2, w\}$ forms a basis for $\mathbb{R}^3$ (c) Express $(2,3,1)^T$ as a linear combination of these basis vectors using orthogonal projections

2. Let U be the subspace of $\mathbb{R}^3$ spanned by $u_1 = (1,1,0)^T$ and $u_2 = (0,1,1)^T$. (a) Find a basis for $U^\perp$ (b) Compute the orthogonal projection of $v = (2,1,3)^T$ onto U (c) Verify that $v - \Pi_U v$ lies in $U^\perp$

3. In the vector space $\mathcal{P}_2$ with inner product $\langle f,g\rangle = \int_0^1 f(x)g(x)\,dx$: (a) Show

that 1 and $x - \frac{1}{2}$ are orthogonal polynomials (b) Use the Gram-Schmidt process to extend these to an orthogonal basis for $\mathcal{P}_2$ (c) Find the orthogonal projection of x^2 onto $\mathrm{span}\{1, x - \frac{1}{2}\}$

4. Let $A = \begin{bmatrix} 1 & 2 & 1 \\ 2 & 1 & 0 \\ 1 & 0 & 1 \end{bmatrix}$. Find: (a) A basis for the column space of A (b) A basis for the null space of A (c) Show these spaces are orthogonal complements by computing appropriate inner products

5. Consider fitting the line $y = mx + b$ to points $(0,1)$, $(1,3)$, and $(2,2)$. (a) Set up the normal equations for this least squares problem (b) Solve for m and b (c) Find the residual vector and verify it is orthogonal to the column space of the coefficient matrix

6. Consider the data points $(1,0)$, $(2,2)$, $(3,1)$, $(4,4)$. Find: (a) The least squares line $y = mx + b$ (b) The orthogonal projection of the data vector onto the column space of the coefficient matrix (c) The residual vector and verify it is orthogonal to both 1 and x

7. Let P be the orthogonal projection onto a subspace U of an inner product space V. Prove that: (a) $P^2 = P$ (b) $P^* = P$ (c) $\|Pv\| \leq \|v\|$ for all $v \in V$

8. Prove that if U and W are orthogonal subspaces of an inner product space V, then:

$$\|u + w\|^2 = \|u\|^2 + \|w\|^2$$

for any $u \in U$ and $w \in W$. Use this to explain why the Pythagorean theorem holds for orthogonal projections.

9. Let V be a finite-dimensional inner product space and $U < V$ a subspace. Prove that: (a) $(U^\perp)^\perp = U$ (b) $\dim U + \dim U^\perp = \dim V$ (c) Every vector in V can be uniquely written as a sum of vectors from U and $U^\perp$

10. For vectors $x, y \in \mathbb{R}^n$ and $\alpha > 0$, consider the regularized least squares problem:

$$\min_{v \in \mathbb{R}^n} \|x - v\|^2 + \alpha \|v - y\|^2$$

(a) Show this has a unique solution (b) Find an explicit formula for the solution in terms of x, y, and α (c) Interpret this solution as an interpolation between x and y

11. Consider the regularized least squares problem:

$$\min_x \|Ax - b\|^2 + \lambda \|x\|^2$$

Prove that as $\lambda \to \infty$, the solution approaches 0, while as $\lambda \to 0^+$, the solution approaches the minimum norm solution of the original least squares problem.

12. For a linear transformation $T : V \to W$ between inner product spaces, prove that $\ker T$ and $\operatorname{im} T^*$ are orthogonal complements in V. What does this tell you about the relationship between solutions to $Tx = b$ and $T^*Tx = T^*b$?

13. Let U be a subspace of an inner product space V. Show that $v \in V$ has minimum norm among all vectors in its coset $v + U$ if and only if $v \perp U$.

14. For a matrix A, show that the orthogonal projection onto $\operatorname{COL}(A)$ is given by $A(A^T A)^{-1} A^T$ when A has full column rank. Use this to explain why the least squares solution minimizes the residual norm.

15. Let V be the space of continuous functions on $[0,1]$ with inner product $\langle f,g \rangle = \int_0^1 f(x)g(x)\,dx$. Show that: (a) The subspace U of functions satisfying $f(0) = f(1)$ is closed under addition and scalar multiplication (b) Find a nonzero function in $U^\perp$ (c) Describe geometrically what it means for a function to be orthogonal to all functions in U

16. Consider training data consisting of hourly temperature readings T and energy usage E (in kWh) from a building's HVAC system:

$$(T,E) = \{(68,42),(72,45),(75,48),(71,44),(69,43),(74,47)\}$$

The facilities manager believes that measurement errors in temperature readings are twice as significant as those in energy readings. Set up and solve an appropriate weighted least squares problem to find the best linear relationship $E = aT + b$.

17. Let A be an $m \times n$ matrix with $m > n$ and let $\boldsymbol{b}$ be a vector such that $A\boldsymbol{x} = \boldsymbol{b}$ has no solution. Prove that among all vectors $\boldsymbol{y}$ satisfying $A\boldsymbol{x} = \boldsymbol{y}$, the vector $\Pi_{\mathrm{im}\,A}\boldsymbol{b}$ is closest to $\boldsymbol{b}$ in Euclidean norm.

18. A manufacturer measures the tensile strength y of a material at different temperatures x, obtaining data points:

$$(x,y) = \{(20,45),(30,42),(40,38),(50,35),(60,30),(70,28)\}$$

Theory suggests the relationship should be of form $y = ae^{bx}$. Show how to transform this into a linear least squares problem and solve for a and b.

19. Let P and Q be orthogonal projections onto subspaces U and W respectively. Show that PQ is itself an orthogonal projection if and only if $PQ = QP$. What does this tell you about the relationship between U and W?

20. An engineer measures signals $s(t)$ contaminated by periodic noise of known frequency ω. The measurements at times $t_1,\ldots,t_n$ are:

$$y_i = s(t_i) + a\cos(\omega t_i) + b\sin(\omega t_i) + \epsilon_i$$

where ϵ_i represents random error. Show how to use orthogonal projections to estimate and remove the periodic components, leaving a cleaner estimate of the original signal $s(t)$.

Chapter 7
Diagonalization & Dynamics

"how is it that all things are chang'd even as in ancient times"

A TRANSFORMATION IN PERPETUAL ACTION reveals patterns hidden from static view. A mass-spring system oscillates with characteristic frequencies; a population grows or declines at intrinsic rates; a network's influence flows along preferred channels. These seemingly distinct phenomena share a common mathematical essence: certain directions remain invariant under repeated transformation, while vectors along these special directions experience pure scaling. Such fixed directions and their associated scaling factors – *eigenvectors* and *eigenvalues* – provide the key to understanding how linear transformations act over time.

Consider a simple linear recurrence modeling population growth: each generation's size is a fixed multiple of the last. The population either grows exponentially or decays to extinction, depending on whether this multiplier exceeds unity. Though elementary, this example contains the seed of a deeper truth: the long-term behavior of linear systems often reduces to pure scaling along special directions. These particular directions and their scaling factors determine not just whether populations thrive or perish, but how mechanical systems vibrate, how heat diffuses, how quantum states evolve, and how networks transmit influence.

The search for such invariant directions leads us to *characteristic polynomials* – equations whose roots reveal the natural scaling factors of a transformation. These *eigenvalues*, together with their associated *eigenvectors*, provide a new perspective on linear transformations. We seek coordinates that aligned with intrinsic scaling; when such exist, the transformation assumes its most primal diagonal form.

This diagonalization – this alignment of coordinates with natural directions – does more than simplify computation. It reveals the essence of how transformations act, decomposing complex motion into simpler components. The vibration of a drum becomes a superposition of pure tones; the flow of heat resolves into independent decay modes. What appears complicated in one basis becomes elementary when seen through the lens of diagonalization.

7.1   The First Order

The simplest differential equation from calculus serves as prototype for all continuous-time linear systems. Consider the equation

$$\frac{dx}{dt} = \lambda x \tag{7.1}$$

Notation: Using the differential operator $D = d/dt$ will prove advantageous later.

where λ is a constant. From calculus, we know the general solution is an exponential:

$$x(t) = x_0 e^{\lambda t}$$

where x_0 is a constant interpretable as the *initial condition* $x_0 = x(0)$. This elementary equation contains the seeds of deeper structure since the solutions to (7.1) are closed under addition and scalar multiplication. As such, the simple solution $e^{\lambda t}$ serves as a basis for the 1-dimensional solution space to (7.1).

Note the centrality of the constant λ in determining the qualitative behavior of all the solutions in the solution space:

Think: The solution corresponding to $x_0 = 0$ is special, in that even when other solutions grow or shrink, it remains constant. Such *equilibrium* solutions are central objects of study in dynamical systems.

- When $\lambda > 0$, solutions grow exponentially
- When $\lambda < 0$, solutions decay exponentially
- When $\lambda = 0$, solutions remain constant

This structure – exponential solutions parametrized by a characteristic number λ – will recur throughout this and the following two chapters. More complex systems will decompose into collections of such fundamental solutions, each growing or decaying at its own characteristic rate. The art lies in finding these natural modes of behavior.

The parameter λ appearing here is the characteristic value that determines the natural rate of growth or decay in the ODE. It is not initially obvious, but such λ values are connected at the deepest levels to the algebra of matrices and linear transformations.

7.2   Coupled First-Order Systems

Real systems rarely evolve in isolation. A predator population depends on its prey; stock prices move with market sectors; neurons fire in vast interconnected networks. Even the simplest model tends to track at least two interacting variables. Consider $x(t)$ and $y(t)$ whose evolution depends linearly on a combination of present states:

$$\frac{dx}{dt} = ax + by \quad : \quad \frac{dy}{dt} = cx + dy$$

where a, b, c, d are constants. Unlike the scalar case, no obvious solution presents itself. Writing this linear system in matrix form reveals a familiar pattern:

$$\frac{d}{dt}\begin{pmatrix} x \\ y \end{pmatrix} = \begin{bmatrix} a & b \\ c & d \end{bmatrix}\begin{pmatrix} x \\ y \end{pmatrix}$$

This matrix form suggests a special case worth examining. Suppose the matrix were diagonal:

$$\frac{d}{dt}\begin{pmatrix} x \\ y \end{pmatrix} = \begin{bmatrix} \lambda_1 & 0 \\ 0 & \lambda_2 \end{bmatrix}\begin{pmatrix} x \\ y \end{pmatrix} \quad \Leftrightarrow \quad \frac{dx}{dt} = \lambda_1 x \ : \ \frac{dy}{dt} = \lambda_2 y$$

each solvable by the methods of the previous section. The solution would be pure exponential growth or decay along each coordinate axis:

$$\begin{pmatrix} x(t) \\ y(t) \end{pmatrix} = \begin{pmatrix} c_1 e^{\lambda_1 t} \\ c_2 e^{\lambda_2 t} \end{pmatrix}$$

where c_1 and c_2 are determined by initial conditions.

This observation – that diagonal systems decompose into independent scalar equations – suggests a strategy. If we could somehow transform our original system into diagonal form, its solution would reduce to pure exponentials: we seek coordinates that reveal the hidden diagonal structure lurking within coupled systems.

Example 7.1 (Chemical Reaction). Consider two chemical species with concentrations x_A and x_B that interact through a simple reaction network in matrix form:

$$\frac{d}{dt}\begin{pmatrix} x_A \\ x_B \end{pmatrix} = \begin{bmatrix} -k_1 & k_2 \\ k_1 & -k_2 \end{bmatrix}\begin{pmatrix} x_A \\ x_B \end{pmatrix}$$

where $k_1, k_2 > 0$ are reaction rate constants. A curious change of coordinates reveals hidden simplicity. Consider the invertible transformation

$$\begin{pmatrix} x_A \\ x_B \end{pmatrix} = \begin{bmatrix} 1 & 1 \\ k_2 & -k_1 \end{bmatrix}\begin{pmatrix} u \\ v \end{pmatrix} \quad \Longleftrightarrow \quad \begin{pmatrix} u \\ v \end{pmatrix} = \begin{bmatrix} 1 & 1 \\ k_2 & -k_1 \end{bmatrix}^{-1}\begin{pmatrix} x_A \\ x_B \end{pmatrix}$$

We can convert the differential equations by means of a similarity transformation:

$$\frac{d}{dt}\begin{pmatrix} u \\ v \end{pmatrix} = \frac{d}{dt}\left(\begin{bmatrix} 1 & 1 \\ k_2 & -k_1 \end{bmatrix}^{-1} \begin{pmatrix} x_A \\ x_B \end{pmatrix} \right) = \begin{bmatrix} 1 & 1 \\ k_2 & -k_1 \end{bmatrix}^{-1} \frac{d}{dt}\begin{pmatrix} x_A \\ x_B \end{pmatrix}$$

$$= \begin{bmatrix} 1 & 1 \\ k_2 & -k_1 \end{bmatrix}^{-1} \begin{bmatrix} -k_1 & k_2 \\ k_1 & -k_2 \end{bmatrix} \begin{pmatrix} x_A \\ x_B \end{pmatrix}$$

$$= \begin{bmatrix} 1 & 1 \\ k_2 & -k_1 \end{bmatrix}^{-1} \begin{bmatrix} -k_1 & k_2 \\ k_1 & -k_2 \end{bmatrix} \begin{bmatrix} 1 & 1 \\ k_2 & -k_1 \end{bmatrix} \begin{pmatrix} u \\ v \end{pmatrix} = \begin{bmatrix} 0 & 0 \\ 0 & -k_1 - k_2 \end{bmatrix} \begin{pmatrix} u \\ v \end{pmatrix}$$

This diagonal matrix has diagonal entries 0 for u and $-(k_1 + k_2) < 0$ for v. Thus u remains constant while v decays exponentially. The solution in original coordinates is computed as:

$$\begin{pmatrix} x_A \\ x_B \end{pmatrix} = \begin{bmatrix} 1 & 1 \\ k_2 & -k_1 \end{bmatrix} \begin{pmatrix} u \\ v \end{pmatrix} = \begin{pmatrix} c_1 + c_2 e^{-(k_1+k_2)t} \\ c_1 k_2 - c_2 k_1 e^{-(k_1+k_2)t} \end{pmatrix}$$

where c_1 and c_2 depend on initial conditions. This "magical" change of coordinates anticipates deeper structure – the zero on the diagonal hints that there is a conservation principle at work (conservation of mass in this case). $\diamond$

The search for such diagonal representations will guide our development in the sections ahead. We shall discover that many coupled systems admit transformation to diagonal form, where their behavior becomes transparent. Those that resist such diagonalization will require more subtle analysis, leading us to the Jordan canonical form of Chapter 8. Throughout, our goal remains to understand how linear systems evolve by finding coordinates that reveal their essential structure.

7.3   Eigenvalues & Eigenvectors

The general linear system of ordinary differential equations takes the form

$$\frac{dx}{dt} = Ax \tag{7.2}$$

where A is a square matrix and $x(t)$ is a vector-valued function of time. Such equations arise throughout engineering – from mechanical vibrations to chemical kinetics to population dynamics. Their solution, though not immediately apparent, holds the key to understanding how linear systems evolve.

Our experience with scalar equations suggests a strategy. Were A replaced by a scalar λ, the solution would take the simple exponential form

$x(t) = e^{\lambda t} x_0$. This observation leads to a crucial question: might there exist a 1-d subspace on which the matrix A acts just like scalar multiplication? This would be equivalent to finding a vector $v \neq 0$ such that

$$Av = \lambda v \tag{7.3}$$

for some scalar λ. Such a special subspace $\mathrm{span}(v)$, should it exist, would be *invariant* under A and, for vectors in this subspace, solving the ODE would reduce to the trivial 1-d case.

Viewing A as a linear transformation illuminates the path forward. Equation (7.3) can be rewritten as

$$(A - \lambda I)v = 0$$

where I denotes the identity matrix. For this equation to have a nonzero solution v, the transformation $(A - \lambda I)$ must have nontrivial kernel. This occurs precisely when $(A - \lambda I)$ fails to be invertible – when its determinant vanishes: $\det(A - \lambda I) = 0$. This is the key to the following fundamental definitions.

Definition 7.2 (Eigenvalues and Eigenvectors). The *characteristic polynomial* $p_A(\lambda)$ of a square matrix A is the polynomial

$$p_A(\lambda) = \det(A - \lambda I) \tag{7.4}$$

A scalar λ is called an *eigenvalue* of A if it is a root of the characteristic polynomial: $p_A(\lambda) = 0$. For each eigenvalue λ, any nonzero vector v satisfying

$$Av = \lambda v \tag{7.5}$$

is called an *eigenvector* corresponding to eigenvalue λ.

Example 7.3. For a concrete example, consider the matrix

$$A = \begin{bmatrix} 2 & 1 \\ 1 & 2 \end{bmatrix}$$

Its characteristic polynomial is

$$\det(A - \lambda I) = \begin{vmatrix} 2 - \lambda & 1 \\ 1 & 2 - \lambda \end{vmatrix} = (2 - \lambda)^2 - 1 = \lambda^2 - 4\lambda + 3$$

Setting this equal to zero yields eigenvalues $\lambda_1 = 3$ and $\lambda_2 = 1$. For $\lambda_1 = 3$, we solve $(A - 3I)v = 0$:

$$\begin{bmatrix} -1 & 1 \\ 1 & -1 \end{bmatrix} \begin{pmatrix} v_1 \\ v_2 \end{pmatrix} = 0 \quad \Rightarrow \quad v_1 = \begin{pmatrix} 1 \\ 1 \end{pmatrix}$$

Nota bene: The restriction $v \neq 0$ is crucial. The zero vector satisfies $A0 = \lambda 0$ for any λ, but tells us nothing about the transformation's behavior.

Nota bene: Eigenvalues and eigenvectors are paired – though uniqueness is not implied.

Similarly, for $\lambda_2 = 1$ we find $v_2 = (1, -1)^T$. Each eigenvector reveals a direction in which A acts by simple scaling: vectors along v_1 are stretched by a factor of 3, while those along v_2 are left unchanged (rescaled by 1). ◇

The search for eigenvalues and eigenvectors thus reduces to:

1. Form the characteristic polynomial $\det(A - \lambda I)$
2. Find its roots (the eigenvalues)
3. For each eigenvalue λ, solve $(A - \lambda I)v = \mathbf{0}$ for nonzero v

As to what type and how many eigenvalues a system has, the following is crucial.

Lemma 7.4 (Characteristic Polynomials). *For any matrix $A \in \mathbb{R}^{n \times n}$, its characteristic polynomial $p_A(\lambda) = \det(A - \lambda I)$ satisfies:*

1. *The polynomial has degree exactly n*
2. *It has exactly n complex roots (counted with algebraic multiplicity)*
3. *Its coefficients are real, and its complex roots occur in conjugate pairs*

Proof. For (1), expand $\det(A - \lambda I)$ using any minor expansion. The term $(-\lambda)^n$ appears uniquely from the product of diagonal entries of $-\lambda I$. All other terms in the determinant expansion contain fewer factors of λ, as they must include at least one entry from A. Thus $p_A(\lambda)$ has degree exactly n with leading coefficient $(-1)^n$.

For (2), the Fundamental Theorem of Algebra ensures that any polynomial of degree n with complex coefficients has exactly n complex roots, counting multiplicity.

For (3), the entries of $A - \lambda I$ are either real constants or real linear terms in λ. The determinant of such a matrix must yield a polynomial with real coefficients. For such polynomials, if $\alpha + i\beta$ is a root, then its complex conjugate $\alpha - i\beta$ must also be a root with equal multiplicity. This follows because complex roots of real polynomials occur in conjugate pairs: if $p(a + bi) = 0$ then $p(\overline{a + bi}) = p(a - bi) = 0$. □

Example: The matrix $J = \begin{bmatrix} 0 & -1 \\ 1 & 0 \end{bmatrix}$ has characteristic polynomial $p_J(\lambda) = \lambda^2 + 1$, whose roots are $\pm i$, illustrating how complex eigenvalues must indeed appear in conjugate pairs. What does the linear transformation with matrix J do?

The characteristic polynomial encodes more than just eigenvalues – its coefficients reveal fundamental invariants of the transformation. Most striking are the relationships between eigenvalues and two elementary matrix measurements: trace and determinant.

Lemma 7.5 (Eigenvalue Relations). *For an $n \times n$ matrix A with eigenvalues $\lambda_1, \ldots, \lambda_n$:*

1. *The determinant equals their product:* $\det(A) = \prod_{i=1}^{n} \lambda_i$
2. *The trace equals their sum:* $tr(A) = \sum_{i=1}^{n} \lambda_i$

Example: For a 2×2 matrix A, the characteristic polynomial $p_A(\lambda) = \lambda^2 - tr(A)\lambda + \det(A)$ makes these relationships transparent.

7.4 Simple Diagonalization

When a linear transformation possesses n distinct real eigenvalues, a remarkable simplification becomes possible. The eigenvectors – those special directions experiencing pure scaling – provide a natural basis for viewing the transformation in its simplest form. This process of changing coordinates to align with eigenvectors, called *diagonalization*, reveals the essential character of the transformation stripped of coupled complexity.

Consider again the matrix from Example 7.3:

$$A = \begin{bmatrix} 2 & 1 \\ 1 & 2 \end{bmatrix}$$

We found eigenvalues $\lambda_1 = 3$ and $\lambda_2 = 1$ with corresponding eigenvectors $v_1 = (1,1)^T$ and $v_2 = (1,-1)^T$. The individual eigenvalue equations

$$Av_1 = 3v_1 \quad \text{and} \quad Av_2 = v_2$$

can be elegantly combined by arranging the eigenvectors as columns of a matrix:

$$A[v_1\ v_2] = [v_1\ v_2] \begin{bmatrix} 3 & 0 \\ 0 & 1 \end{bmatrix}$$

Writing in terms of square matrices $V = [v_1\ v_2]$ and $\Lambda = \text{DIAG}(3,1)$, this becomes simply

$$AV = V\Lambda \tag{7.6}$$

Think: if you have internalized the equation $Av = \lambda v$, you may find the matrix form of this equation to be easily memorable. Just be sure to convince yourself as to the ordering of the terms...

When V is invertible (as it must be when the eigenvalues are distinct), we obtain the diagonalization $V^{-1}AV = \Lambda$. This observation generalizes to arbitrary dimension:

Theorem 7.6 (Diagonalization). *Let A be an $n \times n$ matrix with n distinct real eigenvalues $\lambda_1, \ldots, \lambda_n$ and corresponding eigenvectors $v_1, \ldots, v_n$. Then:*

1. *The eigenvectors form a basis for $\mathbb{R}^n$*
2. *The matrix $V = [v_1 \cdots v_n]$ is invertible*
3. *$V^{-1}AV = \Lambda$ where $\Lambda = \text{DIAG}(\lambda_1, \ldots, \lambda_n)$*

Proof. That the eigenvectors are linearly independent follows from their association with distinct eigenvalues. Indeed, suppose $\sum_{i=1}^{n} c_i v_i = 0$. Then

$$0 = A\left(\sum_{i=1}^{n} c_i v_i\right) = \sum_{i=1}^{n} c_i \lambda_i v_i$$

Subtracting λ_1 times the first equation from the second:

$$\sum_{i=2}^{n} c_i(\lambda_i - \lambda_1)v_i = 0$$

Since $\lambda_i \neq \lambda_1$ for $i \geq 2$, we must have $c_2 = \cdots = c_n = 0$, and consequently $c_1 = 0$. Thus $\{v_1, \ldots, v_n\}$ is linearly independent, making V invertible. The final statement follows from our derivation above. $\qquad\square$

The power of diagonalization lies in how it simplifies computation of matrix powers. When $A = V\Lambda V^{-1}$, repeated multiplication becomes mere diagonal scaling:

Lemma 7.7 (Matrix Powers). *If $A = V\Lambda V^{-1}$ is diagonalizable, then for any positive integer k:*

$$A^k = V\Lambda^k V^{-1}$$

Think: how hard is it to compute powers of the diagonal matrix Λ?

Proof. The result follows directly from associativity of matrix multiplication and the properties of the inverse. $\qquad\square$

This decomposition reveals how a transformation compounds through repeated application: each eigenspace experiences repeated scaling by its eigenvalue, while the change of basis matrices V and V^{-1} translate between our chosen coordinates and these natural eigenspaces. The behavior as $k \to \infty$ becomes transparent – it is controlled entirely by the magnitude of eigenvalues: see Chapter 9 for implications.

The relationship between similar matrices acquires new meaning through diagonalization. Matrices are similar precisely when they represent the same linear transformation viewed in different coordinates. When those coordinates align with an eigenbasis, the matrix assumes diagonal form – its simplest possible representation. Not every matrix admits such diagonal form (a topic for Chapter 8), but when it does, we gain both computational advantage and theoretical insight.

7.5   Matrix Exponentials

Recall from Section 7.1 the general linear system of differential equations:

$$\frac{dx}{dt} = Ax \tag{7.7}$$

where A is a square matrix. The scalar case $A = \lambda I$ yielded solutions of the form $e^{\lambda t} x_0$. Such exponentials indeed generate solutions in the general case. The following definition is key.

Definition 7.8 (Matrix Exponential). The *matrix exponential* of a square matrix A is defined by the absolutely convergent power series

$$e^A = I + A + \frac{A^2}{2!} + \frac{A^3}{3!} + \cdots = \sum_{k=0}^{\infty} \frac{A^k}{k!} \tag{7.8}$$

This formal series inherits many properties of the scalar exponential, including the crucial fact that it solves our differential equation:

Lemma 7.9 (Matrix Exponential Solution). *The general solution to the initial value problem*

$$\frac{dx}{dt} = Ax, \quad x(0) = x_0$$

is given by $x(t) = e^{At}x_0$.

The simplest case occurs when A is diagonal. For $A = \text{DIAG}(\lambda_1, \ldots, \lambda_n)$, the matrix exponential is simply $e^{At} = \text{DIAG}(e^{\lambda_1 t}, \ldots, e^{\lambda_n t})$. Each component evolves independently according to its diagonal entry.

Caveat: While the formula resembles the scalar case, computing e^{At} directly from the series is rarely practical. The art lies in finding more efficient methods based on the structure of A.

Example 7.10 (Diagonal System). The system

$$\frac{d}{dt}\begin{pmatrix} x \\ y \end{pmatrix} = \begin{bmatrix} 2 & 0 \\ 0 & -1 \end{bmatrix}\begin{pmatrix} x \\ y \end{pmatrix}$$

has matrix exponential

$$e^{At} = \begin{bmatrix} e^{2t} & 0 \\ 0 & e^{-t} \end{bmatrix}$$

The first component grows exponentially while the second decays – a behavior determined entirely by the eigenvalues. ◇

Lemma 7.11 (Diagonalizable Matrix Exponential). *If $A = V\Lambda V^{-1}$ is diagonalizable with $\Lambda = \text{DIAG}(\lambda_1, \ldots, \lambda_n)$, then*

$$e^{At} = Ve^{\Lambda t}V^{-1}$$

where $e^{\Lambda t} = \text{DIAG}(e^{\lambda_1 t}, \ldots, e^{\lambda_n t})$.

Proof. By Lemma 7.7, for any positive integer k, $A^k = V\Lambda^k V^{-1}$. Therefore in the power series defining e^{At}:

$$e^{At} = I + At + \frac{(At)^2}{2!} + \cdots = \sum_{k=0}^{\infty} \frac{t^k}{k!}A^k = \sum_{k=0}^{\infty} \frac{t^k}{k!}V\Lambda^k V^{-1}$$

The absolute convergence of the matrix exponential series allows us to factor out V and V^{-1}:

$$e^{At} = V\left(\sum_{k=0}^{\infty} \frac{t^k}{k!}\Lambda^k\right)V^{-1} = Ve^{\Lambda t}V^{-1}$$

where the middle term reduces to $\text{DIAG}(e^{\lambda_1 t}, \ldots, e^{\lambda_n t})$ by the scalar exponential series. □

This decomposition reveals how eigenvalues control the long-term behavior of solutions: each eigendirection experiences exponential growth or decay at a rate determined by its eigenvalue.

Example 7.12 (Population Growth). Consider a simple model of two interacting populations where each species' growth rate depends both on its own population and that of the other species:

$$\frac{d}{dt}\begin{pmatrix} x \\ y \end{pmatrix} = \begin{bmatrix} 2 & 1 \\ 1 & 2 \end{bmatrix}\begin{pmatrix} x \\ y \end{pmatrix}$$

From Example 7.3, we know this matrix has eigenvalues $\lambda_1 = 3$ and $\lambda_2 = 1$ with eigenvectors $v_1 = (1,1)^T$ and $v_2 = (1,-1)^T$. Therefore

$$e^{At} = \begin{bmatrix} 1 & 1 \\ 1 & -1 \end{bmatrix}\begin{bmatrix} e^{3t} & 0 \\ 0 & e^{t} \end{bmatrix}\begin{bmatrix} 1/2 & 1/2 \\ 1/2 & -1/2 \end{bmatrix}$$

The solution reveals two fundamental modes: symmetric growth where both populations increase in proportion (e^{3t} term) and asymmetric growth where their difference evolves more slowly (e^{t} term). Any initial condition excites a combination of these modes, with the symmetric mode eventually dominating due to its larger eigenvalue.   ◇

This connection between eigenvalues and the qualitative behavior of solutions – growth, decay, or oscillation – exemplifies how the algebraic structure of a matrix determines the geometric character of its flow. The matrix exponential transforms our static understanding of eigenvalues into a dynamic picture of how systems evolve in time.

7.6   Higher-Order Equations

A simple mass-spring system requires tracking both position and velocity; a circuit with inductance and capacitance needs current and charge; a chemical reaction network may depend on concentrations and their rates of change. Such systems lead naturally to second-order (or higher) differential equations. Though seemingly more complex than the first-order systems studied thus far, these equations succumb to the same eigenvalue methods through a systematic reduction to matrix form.

As foreshadowed at the conclusion of Chapter 2, consider a linear homogeneous differential equation of order n:

$$\frac{d^n x}{dt^n} + a_{n-1}\frac{d^{n-1}x}{dt^{n-1}} + \cdots + a_1\frac{dx}{dt} + a_0 x = 0$$

where the coefficients a_k are constants. Using the differential operator $D = d/dt$, we can write this more compactly as

$$p(D)x = (D^n + a_{n-1}D^{n-1} + \cdots + a_1 D + a_0 I)x = 0$$

where $p \in \mathcal{P}_n$ is a degree-n polynomial in the differentiation operator D having coefficients a_i, $i = 0\ldots n-1$ (and top coefficient 1). This single equation of order n can be transformed into a system of n first-order equations by introducing new variables for the derivatives. Let

$$x_1 = x, \quad x_2 = \frac{dx}{dt}, \quad x_3 = \frac{d^2x}{dt^2}, \quad \cdots \quad x_n = \frac{d^{n-1}x}{dt^{n-1}}$$

Then our equation becomes a first-order system in matrix form:

$$\frac{d}{dt}\begin{pmatrix} x_1 \\ x_2 \\ x_3 \\ \vdots \\ x_n \end{pmatrix} = \begin{bmatrix} 0 & 1 & 0 & \cdots & 0 \\ 0 & 0 & 1 & \cdots & 0 \\ 0 & 0 & 0 & \cdots & 0 \\ \vdots & \vdots & \vdots & \ddots & \vdots \\ -a_0 & -a_1 & -a_2 & \cdots & -a_{n-1} \end{bmatrix} \begin{pmatrix} x_1 \\ x_2 \\ x_3 \\ \vdots \\ x_n \end{pmatrix}$$

Nota bene: The matrix has a special structure – zeros everywhere except for 1's on the superdiagonal and the coefficients $-a_k$ in the last row. Such matrices are called *companion matrices.*

This transformation reveals a deep connection between the original differential equation and linear algebra: the eigenvalues of the companion matrix are precisely the roots of the *characteristic equation*

$$\lambda^n + a_{n-1}\lambda^{n-1} + \cdots + a_1\lambda + a_0 = p(\lambda) = 0$$

obtained by substituting λ for D in the polynomial operator. When these eigenvalues are distinct (as we assume throughout this chapter), the solution follows directly from our previous work on matrix exponentials.

Example 7.13 (Mass-Spring System). Consider a mass m attached to a spring with constant k and dashpot damping coefficient c. Newton's second law yields

$$m\frac{d^2x}{dt^2} + c\frac{dx}{dt} + kx = 0$$

where $x(t)$ measures displacement from equilibrium. Dividing by m and setting $\omega_0^2 = k/m$ (the natural frequency) and $\gamma = c/m$ (the damping ratio), we obtain

$$\frac{d^2x}{dt^2} + \gamma\frac{dx}{dt} + \omega_0^2 x = 0$$

This transforms to first-order form by setting $x_1 = x$ and $x_2 = \dot{x}$:

$$\frac{d}{dt}\begin{pmatrix} x_1 \\ x_2 \end{pmatrix} = \begin{bmatrix} 0 & 1 \\ -\omega_0^2 & -\gamma \end{bmatrix} \begin{pmatrix} x_1 \\ x_2 \end{pmatrix}$$

For concrete values $\omega_0^2 = 4$ and $\gamma = 3$, the companion matrix

$$A = \begin{bmatrix} 0 & 1 \\ -4 & -3 \end{bmatrix}$$

has characteristic equation $\lambda^2 + 3\lambda + 4 = 0$ with roots $\lambda_1 = -1$ and $\lambda_2 = -2$. The corresponding eigenvectors are

$$v_1 = \begin{pmatrix} 1 \\ -1 \end{pmatrix} \quad \text{and} \quad v_2 = \begin{pmatrix} 1 \\ -2 \end{pmatrix}$$

The solution follows from the matrix exponential. Since A is diagonalizable with

$$V = \begin{bmatrix} 1 & 1 \\ -1 & -2 \end{bmatrix} \quad \text{and} \quad \Lambda = \text{DIAG}(-1, -2) = \begin{bmatrix} -1 & 0 \\ 0 & -2 \end{bmatrix}$$

we have

$$e^{At} = Ve^{\Lambda t}V^{-1} = \begin{bmatrix} 1 & 1 \\ -1 & -2 \end{bmatrix} \begin{bmatrix} e^{-t} & 0 \\ 0 & e^{-2t} \end{bmatrix} \begin{bmatrix} 2 & 1 \\ -1 & -1 \end{bmatrix}$$

For any initial condition $x(0)$, the solution is $x(t) = e^{At}x(0)$. The position $x(t) = x_1(t)$ decays to equilibrium as $t \to \infty$, with the rate determined by the eigenvalues -1 and -2. $\diamond$

This example illustrates how eigenvalue analysis illuminates physical behavior. The eigenvalues -1 and -2 being real and negative indicates pure exponential decay without oscillation – characteristic of an over-damped system. Different parameter values might yield underdamped oscillations or critical damping, cases we shall explore in Chapter 8.

7.7   Basis Solutions

Linear homogeneous differential equations, whether expressed as first-order systems or higher-order scalar equations, admit elegant solution structures based on fundamental modes. These two perspectives – vector-valued solutions to systems and scalar solutions to higher-order equations – offer complementary insights into the nature of linear evolution.

Lemma 7.14 (Solution Spaces). *The solutions to both:*

1. The first-order system $\dfrac{dx}{dt} = Ax$ *in* $\mathbb{R}^n$

2. The n-th order equation $(D^n + a_{n-1}D^{n-1} + \cdots + a_1 D + a_0)x = 0$

form vector spaces under their natural operations of addition and scalar multiplication.

Proof. For the system, if $x_1(t)$ and $x_2(t)$ solve $Dx = Ax$, then:

$$\frac{d}{dt}(x_1 + x_2) = Ax_1 + Ax_2 = A(x_1 + x_2)$$

For the scalar equation, if $x_1(t)$ and $x_2(t)$ solve $p(D)x = 0$ where $p(D)$ is the polynomial differential operator, then:

$$p(D)(x_1 + x_2) = p(D)x_1 + p(D)x_2 = 0$$

Similar reasoning applies to scalar multiplication in both cases. $\qquad\square$

These solution spaces are intimately connected through the eigenstructure of the companion matrix from Section 7.6. The following theorem reveals how eigenvalues generate basis solutions in both settings:

Theorem 7.15 (Basis Solutions). *Let A be diagonalizable with eigenvalues $\lambda_1, \ldots, \lambda_n$. Then:*

1. *For the system $Dx = Ax$ with eigenvectors $v_1, \ldots, v_n$, a basis for the solution space is given by:*
$$\phi_i(t) = e^{\lambda_i t} v_i, \quad i = 1, \ldots, n$$

2. *For the scalar equation $P(D)x = 0$, where $P(\lambda)$ is the characteristic polynomial, a basis is:*
$$\phi_i(t) = e^{\lambda_i t}, \quad i = 1, \ldots, n$$

Proof. For the system, direct substitution verifies each ϕ_i is a solution:

$$\frac{d}{dt}\phi_i = \lambda_i e^{\lambda_i t} v_i = e^{\lambda_i t}(Av_i) = A\phi_i$$

Their linear independence follows from that of the eigenvectors.

For the scalar equation, note that $p(D)e^{\lambda t} = p(\lambda)e^{\lambda t}$. Thus when λ is a root of p, the exponential $e^{\lambda t}$ provides a solution. To establish linear independence of these solutions, suppose we have a linear combination that vanishes:

$$\sum_{i=1}^{n} c_i e^{\lambda_i t} = 0 \quad \text{for all } t$$

We will show all coefficients c_i must be zero. Differentiating this equation k times yields:

$$\sum_{i=1}^{n} c_i \lambda_i^k e^{\lambda_i t} = 0 \quad \text{for } k = 0, 1, \ldots, n-1$$

These n equations form a linear system in the coefficients c_i. The matrix of this system is a Vandermonde matrix in the distinct values λ_i, which has nonzero determinant. Therefore $c_i = 0$ for all i, establishing linear independence of the exponential solutions. $\qquad\square$

Nota bene: A Vandermonde matrix has entries of the form $v_{ij} = x_i^{j-1}$ where x_i are distinct numbers. Its determinant is nonzero precisely when the x_i are distinct. See the Exercises for properties of Vandermonde matrices.

General solutions take parallel forms:

$$x(t) = \sum_{i=1}^{n} c_i e^{\lambda_i t} v_i \quad \text{and} \quad x(t) = \sum_{i=1}^{n} c_i e^{\lambda_i t}$$

The coefficients c_i are determined by initial conditions – either $x(0)$ for the system or $x(0), x'(0), \dots, x^{(n-1)}(0)$ for the scalar equation.

Example 7.16 (Mass-Spring System Redux). The mass-spring equation from Example 7.13:

$$\frac{d^2x}{dt^2} + 3\frac{dx}{dt} + 4x = 0$$

has characteristic equation $\lambda^2 + 3\lambda + 4 = 0$ with roots $\lambda_1 = -1$ and $\lambda_2 = -2$. Thus:

1. As a scalar equation, the general solution is:

$$x(t) = c_1 e^{-t} + c_2 e^{-2t}$$

2. As a system with eigenvectors $v_1 = (1, -1)^T$ and $v_2 = (1, -2)^T$:

$$\begin{pmatrix} x(t) \\ x'(t) \end{pmatrix} = c_1 e^{-t} \begin{pmatrix} 1 \\ -1 \end{pmatrix} + c_2 e^{-2t} \begin{pmatrix} 1 \\ -2 \end{pmatrix}$$

The first component of the vector solution matches the scalar solution, as it must.   ◇

Nota bene: In general, the scalar solutions $\phi_i(t) = e^{\lambda_i t}$ correspond precisely to the first components of the vector solutions $\phi_i(t)$ when the equations are properly matched.

This parallel development reveals the essential unity of linear differential equations, whether viewed as coupled systems or high-order scalar equations. Each perspective offers advantages: the scalar form often simplifies computation when only one variable interests us, while the system form better reveals geometric structure and generalizes to coupled equations.

Foreshadowing: When eigenvalues coincide, both formulations require modification. The solutions acquire polynomial factors multiplying the exponentials – a complication we shall explore in Chapter 8.

The choice between scalar and system forms often depends on context. Physical systems naturally couple multiple variables, favoring the system approach. Signal processing and control theory traditionally use scalar transfer functions, preferring the high-order form. Yet the underlying mathematical structure – exponential solutions built from eigenvalues – remains the same, a unity we shall exploit repeatedly in applications.

• ———————————————————————————— •

Multi-Zone Building Temperature Control

The temperature dynamics of a multi-zone building provide an elegant application of eigenvalue analysis to a practical engineering problem. Consider a building with multiple rooms, each maintained at a controlled temperature through a combination of HVAC input and heat exchange with neighboring spaces. The evolution of temperatures throughout the building reveals the power of linear systems theory in understanding and controlling complex thermal environments.

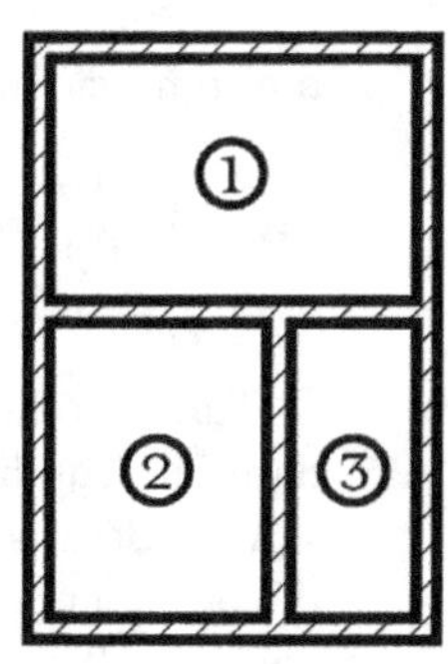

Consider three adjacent rooms sharing walls but with different exterior exposures. Let $T_i(t)$ denote the temperature of room i at time t. The rate of temperature change in each room depends on heat exchange with neighboring rooms (proportional to temperature differences), heat loss to the exterior (proportional to difference from ambient temperature), and HVAC input (controlled heating or cooling).

Applying Newton's law of cooling and conservation of energy leads to a system of coupled differential equations:

$$c_1 \frac{dT_1}{dt} = k_{12}(T_2 - T_1) + k_{13}(T_3 - T_1) - h_1(T_1 - T_a) + u_1$$

$$c_2 \frac{dT_2}{dt} = k_{12}(T_1 - T_2) + k_{23}(T_3 - T_2) - h_2(T_2 - T_a) + u_2$$

$$c_3 \frac{dT_3}{dt} = k_{13}(T_1 - T_3) + k_{23}(T_2 - T_3) - h_3(T_3 - T_a) + u_3$$

where c_i represents the thermal mass of room i, k_{ij} the thermal conductance between rooms i and j, h_i the heat transfer coefficient to ambient temperature T_a, and u_i the HVAC input power to room i.

To analyze this system, first consider the unforced response ($u_i = 0$) relative to ambient temperature. Let $x_i = T_i - T_a$ denote the temperature deviation in room i. Taking realistic values for a modern office building section with equal thermal masses ($c_1 = c_2 = c_3 = 1$), symmetric coupling between adjacent rooms ($k_{12} = k_{23} = 1$, $k_{13} = 0.5$), and varying exterior exposure ($h_1 = 2$, $h_2 = 1$, $h_3 = 1.5$), we obtain the system matrix:

$$A = \begin{bmatrix} -3.5 & 1.0 & 0.5 \\ 1.0 & -2.5 & 1.0 \\ 0.5 & 1.0 & -3.0 \end{bmatrix}$$

The eigenvalues of A determine the natural thermal modes of the building. Computing these reveals three distinct real eigenvalues: $\lambda_1 \approx -4.37$, $\lambda_2 \approx -2.83$, and $\lambda_3 \approx -1.80$. The distinctness of these eigenvalues proves particularly fortunate, as it enables a complete modal decomposition of the system's behavior without requiring the more complex Jordan form analysis developed in Section 7.3. Moreover, their negative values guarantee asymptotic stability - the temperature in each room will eventually return to ambient after any disturbance, as expected from physical intuition.

These eigenmodes reveal the fundamental thermal behavior of the building. The fastest mode, associated with λ_1, represents rapid equilibration between all rooms. The intermediate mode shows temperature oscillation between room 2 and its neighbors, while the slowest mode captures the gradual cooling of the entire system.

Using the diagonalization methods developed in Section 7.5, we can compute

Nota bene: The modeling approach here exemplifies a broader pattern in engineering: complex physical systems often reduce to coupled linear differential equations through appropriate approximations.

Foreshadowing: The appearance of multiple time scales in thermal systems anticipates the more general theory of singular perturbations, crucial in many areas of engineering.

the complete temperature response through the matrix exponential:

$$e^{At} = \begin{bmatrix} -0.65 & -0.24 & 0.72 \\ -0.41 & 0.93 & -0.06 \\ -0.64 & -0.28 & -0.69 \end{bmatrix} \begin{bmatrix} e^{-4.37t} & 0 & 0 \\ 0 & e^{-2.83t} & 0 \\ 0 & 0 & e^{-1.80t} \end{bmatrix} \begin{bmatrix} -0.59 & -0.13 & 0.54 \\ -0.25 & 0.87 & 0.12 \\ -0.52 & -0.38 & -0.83 \end{bmatrix}$$

This decomposition guides practical HVAC control strategy. The fastest modes naturally equilibrate, so control systems should focus on compensating for the slowest decaying mode. Temperature sensors should be placed to best observe the dominant eigenmodes, and building design can optimize thermal response by adjusting the eigenvalue spectrum through insulation and thermal mass distribution.

Example: A well-designed building might have its slowest eigenvalue around -2.0 hr^{-1}, meaning the slowest thermal mode decays with a half-life of about 20 minutes.

The methods developed here extend readily to larger buildings, more complex thermal networks, and other diffusive systems. Whether designing HVAC control systems, optimizing sensor placement, or planning building retrofits, eigenanalysis provides crucial insights into system behavior and guides engineering decisions.

Heavy Vehicle Suspension Analysis

The dynamics of heavily damped suspension systems, such as those found in construction equipment and heavy trucks, provide an instructive application of real eigenvalue analysis. While passenger vehicles typically exhibit oscillatory behavior (a topic for Chapter 8), heavily damped systems demonstrate pure exponential modes that perfectly align with the theory developed in this chapter.

Consider a "quarter-truck" model of one wheel on a heavy vehicle. Let $x(t)$ denote the vertical displacement of the vehicle body (sprung mass) and $y(t)$ the displacement of the wheel assembly (unsprung mass) from their equilibrium positions. The system obeys Newton's second law applied to each mass:

$$M\ddot{x} = -k(x - y) - c(\dot{x} - \dot{y})$$

$$m\ddot{y} = k(x - y) + c(\dot{x} - \dot{y}) - k_t y$$

where M is the quarter-truck body mass, m the wheel assembly mass, k the suspension spring constant, c the damper coefficient, and k_t the tire spring rate.

To analyze this fourth-order system through eigentheory, we transform it to first-order form by introducing the state vector:

$$z = \begin{pmatrix} x \\ y \\ \dot{x} \\ \dot{y} \end{pmatrix}$$

This yields the matrix equation $\dot{z} = Az$ where:

$$A = \begin{bmatrix} 0 & 0 & 1 & 0 \\ 0 & 0 & 0 & 1 \\ -k/M & k/M & -c/M & c/M \\ k/m & -(k+k_t)/m & c/m & -c/m \end{bmatrix}$$

Taking values typical for a heavy commercial vehicle ($M = 2500$ kg, $m = 200$ kg, $k = 100000$ N/m, $k_t = 800000$ N/m, $c = 15000$ Ns/m), the eigenvalues emerge as four distinct real values:

$$\lambda_1 = -12.3, \quad \lambda_2 = -8.4, \quad \lambda_3 = -3.2, \quad \lambda_4 = -0.9$$

Think: The wide separation between eigenvalues reflects different physical time scales: rapid tire deflection, intermediate suspension motion, and slow body settling.

These distinct real eigenvalues, characteristic of overdamped systems, indicate pure exponential decay modes without oscillation. The explicit solution through matrix exponential takes the form developed in Section 7.5:

$$e^{At} = \sum_{j=1}^{4} e^{\lambda_j t} v_j w_j^T$$

where v_j and w_j^T are the right and left eigenvectors respectively. Each mode decays at its own characteristic rate, with physical interpretations:

1. λ_1 mode: rapid tire deflection absorption
2. λ_2 mode: primary suspension response
3. λ_3 mode: coupled body-suspension motion
4. λ_4 mode: slow final settling

This eigenstructure guides heavy vehicle suspension design through several principles:

1. All eigenvalues should be real and negative for pure damping
2. Time scale separation prevents undesirable coupling
3. Mode shapes (eigenvectors) determine effective sensor placement
4. Slowest mode (λ_4) governs overall settling time

The design of construction equipment often deliberately seeks this overdamped behavior to maintain stable operation under varying loads. The eigenvectors determine optimal placement of sensors and actuators by revealing how each decay mode manifests in measurable motions.

Example: Mining trucks use multiple displacement sensors to monitor suspension motion, with sensor locations chosen to best observe the dominant eigenmodes identified through this analysis.

Extension to full-vehicle dynamics introduces additional degrees of freedom - vertical motion at each corner plus pitch and roll of the body - leading to larger matrices but similar principles. The eigenstructure reveals coupled modes like load-sharing between diagonally opposite corners that prove crucial for stability under shifting loads.

This mechanical application complements our thermal analysis by demonstrating eigentheory's power for a different class of physical system. Where thermal modes showed simple exponential decay from geometric symmetry, suspension modes exhibit a hierarchy of decay rates from mechanical coupling. Together they illustrate how the mathematical framework developed in this chapter illuminates diverse physical phenomena, transforming fourth-order differential equations into clearly understood modal responses through the unifying language of linear algebra.

Exercises: Chapter 7

1. Let $A = \begin{bmatrix} 3 & 1 \\ 1 & 3 \end{bmatrix}$. Find its eigenvalues and eigenvectors. Use these to compute e^{At} and describe the behavior of solutions to the system $\dot{x} = Ax$ as $t \to \infty$.

2. Find the eigenvalues and eigenvectors of $A = \begin{bmatrix} 4 & -1 \\ 2 & 1 \end{bmatrix}$. Use these to solve the initial value problem $\dot{x} = Ax$ with $x(0) = \begin{pmatrix} 1 \\ 0 \end{pmatrix}$.

3. Compute the eigenvalues and eigenvectors of $B = \begin{bmatrix} 3 & -2 & 0 \\ 0 & 3 & 0 \\ -4 & 1 & 1 \end{bmatrix}$. Is B diagonalizable?

4. Prove that if A is a triangular matrix (upper or lower) then the diagonal entries coincide with the eigenvalues.

   This is an extremely useful result.

5. Convert the equation $\ddot{x} + 2\dot{x} + 5x = 0$ to a first-order system using the companion matrix. Show the relationship between the eigenvalues of this matrix and the roots of the characteristic equation of the original second-order equation.

6. For the differential equation $\ddot{x} + 3\dot{x} + 2x = 0$, find all linearly independent solutions of the form $e^{\lambda t}$. Write out the general solution using arbitrary constants c_i.

7. Let $A = \begin{bmatrix} 1 & 1 \\ 0 & 1 \end{bmatrix}$ and compute e^{At} by summing the first four terms of its series expansion. Compare this with the result obtained by diagonalization. Why do these differ?

8. A matrix A has characteristic polynomial $p(\lambda) = \lambda^3 - 6\lambda^2 + 11\lambda - 6$. Is A diagonalizable?

9. A matrix A has characteristic polynomial $p(\lambda) = \lambda^4 - 2\lambda^3 - \lambda^2 + 2\lambda$. Find its eigenvalues and determine whether A is necessarily diagonalizable. What can you say about the long-term behavior of solutions to $\dot{x} = Ax$?

10. The trace of a matrix A is the sum of its diagonal entries. Prove that the trace equals the sum of the eigenvalues (counted with multiplicity).

11. Consider the third-order equation $\dddot{x} + 4\ddot{x} + 5\dot{x} + 2x = 0$. Find a basis for its solution space and express the general solution. For what values of the initial conditions $x(0)$, $\dot{x}(0)$, and $\ddot{x}(0)$ does the solution remain bounded as $t \to \infty$?

12. Show that for any diagonalizable matrix A, the matrices A and e^A share the same eigenvectors. If v is an eigenvector of A with eigenvalue λ, what is the corresponding eigenvalue of e^A?

13. Prove that the determinant of a square matrix A equals the product of its eigenvalues: for simplicity, assume that A is diagonalizable, though the result holds in general.

14. Show that if A is diagonalizable with all eigenvalues real and negative, then $\lim_{t \to \infty} e^{At} = 0$ (the zero matrix).

15. If A is a diagonalizable 2×2 matrix and $\operatorname{tr}(A) = 0$, what can you say about its eigenvalues? What does this tell you about solutions to the system $\dot{x} = Ax$?

16. Consider square n-by-n matrices A and B. What condition is required to con-

clude the innocuous-looking result $e^{(A+B)t} = e^{At}e^{Bt}$. Give an example in the 2-by-2 case for which this formula does not hold.

17. Consider the matrix $A = \begin{bmatrix} 3 & -1 \\ 2 & 4 \end{bmatrix}$. Compute its characteristic polynomial $p(\lambda)$ and verify directly that $p(A) = 0$ (the zero matrix). What does this suggest about the relationship between a matrix and its characteristic polynomial?

18. The *Cayley-Hamilton theorem* states that every square matrix satisfies its own characteristic equation. For $A = \begin{bmatrix} 2 & 1 \\ 0 & 2 \end{bmatrix}$, verify this by showing that if $p(\lambda) = \det(A - \lambda I)$ then $p(A) = 0$. What does this tell you about powers of $A - 2I$?

19. For any 2×2 matrix A, show that $A^2 - \text{tr}(A)A + \det(A)I = 0$. Explain how this is a special case of Cayley-Hamilton for 2×2 matrices.

20. Suppose A is a 3×3 matrix with characteristic polynomial $p(\lambda) = \lambda^3 - 7\lambda^2 + 14\lambda - 8$. Without computing any matrix operations, determine the trace and determinant of A, and give an expression for A^3 in terms of lower powers of A

21. A rotation matrix R on $\mathbb{R}^3$ is an orthogonal matrix with has determinant 1. Explain why R must have 1 as an eigenvalue, thus rotating about some fixed axis. Does this argument extend to rotations in $\mathbb{R}^n$ for all n? (Hint: consider the characteristic polynomial and what happens in odd versus even dimensions).

22. Prove that for any square matrix A, the eigenvalues of e^A are $\{e^\lambda\}$ where $\{\lambda\}$ are the eigenvalues of A. Explain why this means e^A is always invertible.

23. The previous exercise shows that for A a square matrix, e^A is invertible. What is the inverse? Was it easier to guess the inverse than it was to show that it is invertible?

24. The Cayley-Hamilton theorem provides a path to computing the inverse of a nonsingular matrix. If $p(\lambda)$ is the characteristic polynomial of A, write $p(\lambda) = a_n\lambda^n + \cdots + a_1\lambda + a_0$ and show that when A is invertible:

$$A^{-1} = -\frac{1}{a_0}(a_n A^{n-1} + \cdots + a_2 A + a_1 I)$$

Use this to find A^{-1} for $A = \begin{bmatrix} 3 & 1 \\ 1 & 3 \end{bmatrix}$ and compare with other methods.

25. For a 2×2 matrix A with distinct eigenvalues λ_1, λ_2, use the Cayley-Hamilton theorem to prove that:

$$A = \frac{\lambda_2 I - A}{\lambda_2 - \lambda_1}\lambda_1 + \frac{A - \lambda_1 I}{\lambda_2 - \lambda_1}\lambda_2$$

This expression shows how A decomposes along its eigenspaces even without explicitly computing eigenvectors.

26. For the 3×3 Vandermonde matrix with $x_1 = 1$, $x_2 = 2$, and $x_3 = 3$, find its eigenvalues and verify they are real. Then compute the trace and verify it equals $x_1 + x_2 + x_3$ plus additional terms you should determine. What does this suggest about the general relationship between the x_i values and the trace of a Vandermonde matrix?

27. For distinct real numbers $x_1, \ldots, x_n$, the Vandermonde matrix V is defined by

$v_{ij} = (x_i)^{j-1}$:

$$V = \begin{bmatrix} 1 & x_1 & x_1^2 & \cdots & x_1^{n-1} \\ 1 & x_2 & x_2^2 & \cdots & x_2^{n-1} \\ \vdots & \vdots & \vdots & \ddots & \vdots \\ 1 & x_n & x_n^2 & \cdots & x_n^{n-1} \end{bmatrix}$$

Show first that when $n = 2$, $\det(V) = x_2 - x_1$. Then prove for $n = 3$ that $\det(V) = (x_2 - x_1)(x_3 - x_1)(x_3 - x_2)$. Based on this pattern, state a conjecture for the general formula $\det(V)$ in terms of differences of the x_i. Prove this formula. What does this tell you about the eigenvalues when $x_i = i$?

Chapter 8
Eigenvalue Complexities

"With songs of sweetest cadence to the turning spindle & reel"

REALITY TRANSCENDS REAL EXPONENTIALS. The elegant theory of the previous chapter – with its real eigenvalues generating pure growth and decay – must confront a deeper truth: the world pulses with rhythm. Heart cells oscillate in synchronized waves; animal populations surge and collapse in predictable cycles; financial markets breathe between boom and bust. Even our thoughts and social movements follow patterns of ebb and flow, as ideas fade and resurge with generations. This ubiquitous periodicity seems to defy our carefully constructed framework of eigenvalues and exponential solutions.

Yet mathematics bends rather than breaks. The introduction of complex eigenvalues resolves our crisis of description, revealing oscillation as natural as growth or decay. When eigenvalues coincide, still richer patterns emerge – polynomial terms multiplying our exponentials in ways that mirror the gradual onset of synchronized behavior in coupled systems. What appears first as complications to our theory reveals itself as essential structure, demanded by the very phenomena we seek to understand.

These complexities – imaginary eigenvalues, repeated roots, and their subtle interactions – lead us beyond simple diagonalization to the best possible world of the Jordan Canonical Form. This framework unifies our treatment of linear evolution while illuminating the delicate ways in which nearly identical eigenvalues can produce radically different behavior. The mathematics itself seems to pulse between poles of simplicity and intricacy, echoing the rhythms it describes.

8.1   Complex Eigenvalues & Oscillation

The simple harmonic oscillator – a mass on a spring without damping – provides our first encounter with behavior that transcends real eigenvalues. The equation of motion

$$\frac{d^2 x}{dt^2} + \omega^2 x = 0$$

has well-known solutions involving sines and cosines:

$$x(t) = c_1 \cos(\omega t) + c_2 \sin(\omega t)$$

Yet these solutions seem to conflict with our theory from Chapter 7, where all solutions emerged as combinations of real exponentials. Converting to first-order form via $x_1 = x$ and $x_2 = \dot{x}$ yields

$$\frac{d}{dt} \begin{pmatrix} x_1 \\ x_2 \end{pmatrix} = \begin{bmatrix} 0 & 1 \\ -\omega^2 & 0 \end{bmatrix} \begin{pmatrix} x_1 \\ x_2 \end{pmatrix}$$

The characteristic polynomial $\det(A - \lambda I) = \lambda^2 + \omega^2$ has roots $\lambda = \pm i\omega$ – our first encounter with complex eigenvalues.

Example 8.1 (Complex Eigenvectors). Consider the rotation matrix for angle $\pi/3$:

$$R = \begin{bmatrix} \cos(\pi/3) & -\sin(\pi/3) \\ \sin(\pi/3) & \cos(\pi/3) \end{bmatrix} = \begin{bmatrix} 1/2 & -\sqrt{3}/2 \\ \sqrt{3}/2 & 1/2 \end{bmatrix}$$

The characteristic equation is $\lambda^2 - \lambda + 1 = 0$, yielding eigenvalues $\lambda = \frac{1}{2} \pm i\frac{\sqrt{3}}{2} = e^{\pm i\pi/3}$. For $\lambda = \frac{1}{2} + i\frac{\sqrt{3}}{2}$, we solve $(R - \lambda I)v = \mathbf{0}$:

$$-\frac{\sqrt{3}}{2} \begin{bmatrix} i & 1 \\ -1 & i \end{bmatrix} \begin{pmatrix} v_1 \\ v_2 \end{pmatrix} = \mathbf{0}$$

yielding eigenvector $v = (1, i)^T$. Though this eigenvector is complex, the matrix R maps real vectors to real vectors. The seeming paradox resolves by noting that v and its complex conjugate $\bar{v} = (1, -i)^T$ span $\mathbb{R}^2$ over the complex numbers, while their real and imaginary parts span $\mathbb{R}^2$ over the reals. ◇

Far from invalidating our theory, these complex eigenvalues illuminate it. Euler's formula $e^{i\theta} = \cos\theta + i\sin\theta$ reveals that our trigonometric solutions are simply complex exponentials in disguise:

$$c_1 \cos(\omega t) + c_2 \sin(\omega t) = \Re\left(z_1 e^{i\omega t} + z_2 e^{-i\omega t} \right)$$

for appropriate complex constants z_1 and z_2. The oscillatory motion emerges from exponential solutions with purely imaginary exponents.

This insight generalizes through study of the fundamental 2×2 matrix

$$J = \begin{bmatrix} 0 & -1 \\ 1 & 0 \end{bmatrix} \tag{8.1}$$

Nota bene: The matrix $J = \sqrt{-I}$ is the matrix analogue of i in complex arithmetic.

Direct computation reveals that $J^2 = -I$, suggesting a deep connection to complex arithmetic. Indeed, the matrix exponential e^{Jt} can be computed directly from its series definition:

$$
\begin{aligned}
e^{Jt} &= I + tJ + \tfrac{t^2}{2!}J^2 + \tfrac{t^3}{3!}J^3 + \tfrac{t^4}{4!}J^4 + \cdots \\
&= \left(I - \tfrac{t^2}{2!}I + \tfrac{t^4}{4!}I - \cdots\right) + \left(tJ - \tfrac{t^3}{3!}J + \tfrac{t^5}{5!}J - \cdots\right) \tag{8.2} \\
&= (\cos t)I + (\sin t)J
\end{aligned}
$$

This matrix-valued analog of Euler's formula explains the geometric action of J – pure rotation in the plane.

More generally, consider a matrix A with complex conjugate eigenvalues $\alpha \pm i\beta$. This matrix – like J – cannot be diagonalized over the reals, but we can reduce it to a simplest form.

Such matrices arise frequently in applications – the real part α controlling growth or decay while the imaginary part β determines oscillation frequency.

Lemma 8.2 (Complex Normal Form). *Any 2×2 matrix A with complex conjugate eigenvalues $\alpha \pm i\beta$ is similar to $\alpha I + \beta J$. That is, there exists an invertible matrix P such that*

$$P^{-1}AP = \alpha I + \beta J \tag{8.3}$$

Proof. Let $v = u + iw$ be an eigenvector for eigenvalue $\alpha + i\beta$. Then

$$Av = (\alpha + i\beta)v \implies A(u + iw) = (\alpha + i\beta)(u + iw)$$

Equating real and imaginary parts:

$$Au = \alpha u - \beta w \quad \text{and} \quad Aw = \beta u + \alpha w$$

The matrix $P = [u \ w]$ is invertible (as v cannot be purely real or imaginary), and

$$AP = [\alpha u - \beta w \ \ \beta u + \alpha w] = P(\alpha I + \beta J)$$

yielding the desired similarity. $\square$

This *normal form* is the simplest possible presentation of a 2-by-2 matrix with complex eigenvalues. Based on how similar matrices behave under exponentiation, our strategy for exponentiating A is to change coordinates, exponentiate $\alpha I + \beta J$, then change coordinates back. Euler's Theorem once again emerges.

Lemma 8.3 (Euler's Theorem Redux).

$$e^{(\alpha I + \beta J)t} = e^{\alpha t}(\cos(\beta t)I + \sin(\beta t)J) \tag{8.4}$$

Proof. Since I and J commute, we have:

$$e^{(\alpha I + \beta J)t} = e^{\alpha t I + \beta t J} = e^{\alpha t I}e^{\beta t J}$$

Having computed e^{Jt} in Equation (8.2), we conclude:

$$e^{(\alpha I + \beta J)t} = e^{\alpha t}I(\cos(\beta t)I + \sin(\beta t)J) = e^{\alpha t}(\cos(\beta t)I + \sin(\beta t)J)$$

as claimed. $\qquad\square$

This decomposition reveals how complex eigenvalues generate spiraling motion – exponential growth or decay modulated by rotation.

8.2   Repeated Eigenvalues

Not all transformations keep their eigenspaces distinct. When eigenvalues coincide, the corresponding eigenvectors may merge or proliferate in subtle ways. This collision of eigenspaces – this failure of perfect separation – leads to behavior richer than pure scaling, yet still simpler than complex oscillation. Understanding such behavior requires carefully distinguishing between how often an eigenvalue appears and how many independent directions it controls.

Consider the matrices

$$A_1 = \begin{bmatrix} 2 & 0 \\ 0 & 2 \end{bmatrix} \quad \text{and} \quad A_2 = \begin{bmatrix} 2 & 1 \\ 0 & 2 \end{bmatrix}$$

Both have characteristic polynomial $(\lambda - 2)^2$, yielding eigenvalue $\lambda = 2$ with *algebraic multiplicity* 2. Yet these matrices behave quite differently: A_1 acts by pure scaling in all directions, while A_2 combines scaling with shear. This distinction emerges from the *geometric multiplicity* – the dimension of the eigenspace $\ker(A - 2I)$.

Definition 8.4 (Eigenvalue Multiplicities). The *algebraic multiplicity* of an eigenvalue is its multiplicity as a root of the characteristic polynomial. The *geometric multiplicity* is the dimension of its eigenspace $\ker(A - \lambda I)$. •

The geometric multiplicity never exceeds the algebraic multiplicity.

For A_1 above, both multiplicities equal 2 – every nonzero vector is an eigenvector with eigenvalue 2. For A_2, the algebraic multiplicity is 2 but the geometric multiplicity is only 1, as only multiples of $(1, 0)^T$ are eigenvectors. This deficiency of eigenvectors signals deeper structure that pure diagonalization cannot capture.

When geometric and algebraic multiplicities differ, eigenvectors alone cannot span the space. We require *generalized eigenvectors* – vectors w satisfying $(A - \lambda I)^k w = 0$ for some $k > 1$. These vectors generate solutions involving polynomial terms multiplied by exponentials, as seen in our example above.

Lemma 8.5 (Generalized Eigenspaces). *For eigenvalue λ, the sequence of subspaces*

$$\ker(A - \lambda I) < \ker(A - \lambda I)^2 < \ker(A - \lambda I)^3 < \cdots$$

stabilizes at dimension equal to the algebraic multiplicity of λ.

Example 8.6 (Generalized Eigenvectors). For the matrix

$$A = \begin{bmatrix} 2 & 1 & 0 \\ 0 & 2 & 1 \\ 0 & 0 & 2 \end{bmatrix}$$

the eigenvalue $\lambda = 2$ has algebraic multiplicity 3 but geometric multiplicity 1. The chain of generalized eigenvectors

$$w_1 = \begin{pmatrix} 1 \\ 0 \\ 0 \end{pmatrix}, \quad w_2 = \begin{pmatrix} 0 \\ 1 \\ 0 \end{pmatrix}, \quad w_3 = \begin{pmatrix} 0 \\ 0 \\ 1 \end{pmatrix}$$

satisfies

$$(A - 2I)w_1 = 0, \quad (A - 2I)w_2 = w_1, \quad (A - 2I)w_3 = w_2$$

$\diamond$

BONUS! This nesting of generalized eigenspaces – or any sequence of nested subspaces – is called a *filtration*. Filtrations are important in approximating large complicated spaces by graded low-dimensional entities, *cf.* Taylor polynomials, Fourier series, etc.

This richer structure – generalized eigenvectors forming chains of increasing length – provides the key to understanding transformations with repeated eigenvalues. Though such transformations resist diagonalization, their behavior remains comprehensible through careful analysis of how the generalized eigenspaces interact. This understanding will prove crucial as we develop the complete theory of linear evolution in subsequent sections.

Example 8.7 (Exponentiating a Simple Block). Consider the matrix

$$R = \begin{bmatrix} 2 & 1 & 0 \\ 0 & 2 & 1 \\ 0 & 0 & 2 \end{bmatrix}$$

This matrix can be written as $R = 2I + N$ where I is the identity matrix and

$$N = \begin{bmatrix} 0 & 1 & 0 \\ 0 & 0 & 1 \\ 0 & 0 & 0 \end{bmatrix}$$

is *nilpotent* – meaning some power of N equals zero. Indeed, $N^3 = 0$ in this case, though $N^2 \neq 0$. To find e^{Rt}, we use an exponent rule:

$$e^{Rt} = e^{(2I+N)t} = e^{2It}e^{Nt} = e^{2t}e^{Nt}$$

this being valid since I commutes with any matrix. The challenge of computing e^{Nt} is not so bad, thanks to the exponential series and nilpotency:

$$\begin{aligned} e^{Nt} &= I + Nt + \frac{N^2 t^2}{2!} + \frac{N^3 t^3}{3!} + \cdots \\ &= I + Nt + \frac{N^2 t^2}{2!} \end{aligned}$$

The series terminates since all higher powers of N vanish. Computing N^2 and evaluating yields:

$$e^{Rt} = e^{2t}e^{Nt} = e^{2t}\begin{bmatrix} 1 & t & \frac{t^2}{2} \\ 0 & 1 & t \\ 0 & 0 & 1 \end{bmatrix} = \begin{bmatrix} e^{2t} & te^{2t} & \frac{t^2}{2}e^{2t} \\ 0 & e^{2t} & te^{2t} \\ 0 & 0 & e^{2t} \end{bmatrix}$$

The solution reveals a remarkable structure: the exponential growth e^{2t} is modulated by polynomial terms in t. This pattern – polynomial-exponential products – characterizes solutions to systems with repeated eigenvalues. The nilpotent part N creates a cascade of influence up the superdiagonal, with each level inheriting the dynamics of those below it.

◇

8.3   *The Jordan Canonical Form*

The quest for simplest form leads us beyond diagonalization. When eigenvalues coincide or turn complex, pure scaling gives way to richer structure – a cascade of influence captured by the *Jordan canonical form*. This ultimate decomposition reveals the true anatomy of linear transformations, showing how general evolution emerges from an interplay of scaling, rotation, and generalized eigenvectors.

Our journey through repeated eigenvalues and polynomial-exponential solutions points toward a deeper unity. Recall that matrices with repeated eigenvalues may lack a full set of independent eigenvectors, requiring generalized eigenvectors to span the space. The previous example showed how such structure manifests in solutions – pure exponentials gain polynomial coefficients through a cascade of influence along chains of generalized eigenvectors. These patterns, seemingly special cases, in fact reveal the general form all linear transformations must take.

Theorem 8.8 (Jordan Canonical Form). *Every square matrix A is similar to a block diagonal matrix J, called its* Jordan canonical form:

$$A = VJV^{-1} = V \begin{bmatrix} J_1 & 0 & \cdots & 0 \\ 0 & J_2 & \cdots & 0 \\ \vdots & \vdots & \ddots & \vdots \\ 0 & 0 & \cdots & J_m \end{bmatrix} V^{-1}$$

where each Jordan block J_i *has one of two forms:*

1. *For a real eigenvalue λ of algebraic multiplicity k, the Jordan block is the k-by-k matrix:*

$$J_i = \lambda I + N = \begin{bmatrix} \lambda & 1 & 0 & \cdots & 0 \\ 0 & \lambda & 1 & \cdots & 0 \\ 0 & 0 & \lambda & \ddots & \vdots \\ \vdots & \vdots & \ddots & \ddots & 1 \\ 0 & 0 & \cdots & 0 & \lambda \end{bmatrix}$$

2. *For complex conjugate eigenvalues $\alpha \pm i\beta$ of algebraic muliplicity k, the Jordan block is the $2k$-by-$2k$ block matrix*

$$J_i = \begin{bmatrix} C & I & 0 & \cdots & 0 \\ 0 & C & I & \cdots & 0 \\ 0 & 0 & C & \ddots & \vdots \\ \vdots & \vdots & \ddots & \ddots & I \\ 0 & 0 & \cdots & 0 & C \end{bmatrix} \quad : \quad C = \begin{bmatrix} \alpha & -\beta \\ \beta & \alpha \end{bmatrix} \quad : \quad I = \begin{bmatrix} 1 & 0 \\ 0 & 1 \end{bmatrix}$$

   where each entry is a 2-by-2 block.

The form J is unique up to the ordering of blocks.

Each Jordan block corresponds to an eigenvalue (real or complex) and its associated generalized eigenvectors. The size of the block equals the length of the longest chain of generalized eigenvectors for that eigenvalue. When all eigenvalues are real and distinct (or real but with independent eigenvectors), each block is 1×1 and we recover the diagonal form of Chapter 7.

This block structure mirrors that of the real case, with the superdiagonal I matrices playing the role of the nilpotent matrix of ones, connecting a chain of 2×2 complex blocks C.

It would have been simpler to use the complex-valued Jordan form; only one type of block is needed. Given our motivation in working with solutions to ODEs, we have chosen to keep things real.

The structure within each block illuminates the transformation's action. For real eigenvalues, the matrix N having ones on the superdiagonal and zeros elsewhere is *nilpotent* – some power equals zero. This nilpotence explains the finite polynomial terms we saw emerge in solutions. For complex eigenvalues, the 2×2 blocks encode rotation as seen in Section 8.1.

Example 8.9 (Jordan Structure). The matrix

$$A = \begin{bmatrix} 2 & 1 & 0 & 0 \\ 0 & 2 & 0 & 0 \\ 0 & 0 & 3 & 1 \\ 0 & 0 & 0 & 3 \end{bmatrix}$$

is already in Jordan form. It has two Jordan blocks: a 2×2 block for eigenvalue $\lambda = 2$ and another for $\lambda = 3$. The generalized eigenvector chains have lengths 2 and 2 respectively. Solutions will involve terms like te^{2t} and te^{3t} from each block's nilpotent part.   ◇

The Jordan decomposition provides the key to understanding matrix powers and exponentials. Since J is block diagonal, its exponential is also block diagonal:

$$e^{At} = Ve^{Jt}V^{-1} = V \begin{bmatrix} e^{J_1 t} & 0 & \cdots & 0 \\ 0 & e^{J_2 t} & \cdots & 0 \\ \vdots & \vdots & \ddots & \vdots \\ 0 & 0 & \cdots & e^{J_k t} \end{bmatrix} V^{-1}$$

For each real Jordan block $J_i = \lambda I + N$, we have:

$$e^{J_i t} = e^{\lambda t} e^{Nt} = e^{\lambda t} \left(I + Nt + \frac{N^2 t^2}{2!} + \cdots + \frac{N^{m-1} t^{m-1}}{(m-1)!} \right)$$

where m is the size of the block. The series terminates because $N^m = 0$. For complex blocks, similar formulas emerge using the rotation matrices studied earlier.

Example 8.10 (Full Jordan Decomposition). Consider the matrix

$$A = \begin{bmatrix} 7 & -4 & 4 \\ 4 & -1 & 4 \\ 0 & 0 & 3 \end{bmatrix}$$

Its characteristic polynomial $(\lambda - 3)(\lambda - 3)(\lambda - 3)$ reveals eigenvalue $\lambda = 3$ with algebraic multiplicity 3. Computing generalized eigenvectors shows geometric multiplicity 1, yielding Jordan form:

$$A = V \begin{bmatrix} 3 & 1 & 0 \\ 0 & 3 & 1 \\ 0 & 0 & 3 \end{bmatrix} V^{-1}$$

where V contains the generalized eigenvectors. The single Jordan block indicates all generalized eigenvectors form one chain. Solutions will involve terms e^{3t}, te^{3t}, and $t^2 e^{3t}/2$.

◇

Nota bene: computing the generalized eigenvector that effect the coordinate transformation V is not always straightforward. See Example 8.11 for details.

The Jordan form reveals that every linear transformation, viewed in the right coordinates, acts through a combination of:

1. Pure scaling (diagonal entries)
2. Rotation (complex conjugate blocks)
3. Cascading influence (superdiagonal ones)

This decomposition provides the complete picture of linear evolution, unifying our earlier studies of eigenvalues, repeated roots, and complex behavior into a single coherent structure.

8.4   Finding the Jordan Form

The beautiful structure of Jordan form revealed in previous sections now confronts a practical challenge: given a matrix A, how does one actually compute the similarity transformation V putting A into Jordan form? Unlike eigenvalues, which emerge cleanly from characteristic polynomials, or eigenvectors found through nullspace computation, the Jordan structure demands careful investigation of how generalized eigenvectors chain together into a basis. Our task is to transform theoretical understanding into practical computation.

The process begins with eigenvalues and their multiplicities. For each eigenvalue λ:

1. The algebraic multiplicity m comes from the characteristic polynomial
2. The geometric multiplicity $k = \dim \ker(A - \lambda I)$ counts independent eigenvectors
3. The difference $m - k$ reveals how many generalized eigenvectors we need

Yet these numbers alone tell only part of the story. The real work lies in discovering how generalized eigenvectors form chains that build our transformation matrix V.

Recall from Section 8.2 that each eigenvalue λ generates a sequence of nested subspaces:

$$\ker(A - \lambda I) < \ker(A - \lambda I)^2 < \ker(A - \lambda I)^3 < \cdots$$

This filtration stabilizes when it reaches dimension equal to λ's algebraic multiplicity. The challenge lies in extracting chains of generalized eigenvectors from this nested structure. Each chain begins with an eigenvector v_1 and builds through solving equations of form:

$$(A - \lambda I)v_{i+1} = v_i$$

When multiple chains exist, we must carefully track their relationships to construct V properly.

Example 8.11 (Jordan Transformation). Consider the 5×5 matrix:

$$A = \begin{bmatrix} 3 & 2 & 1 & 0 & 2 \\ 0 & 3 & -1 & 0 & 3 \\ 0 & 0 & 3 & 0 & -4 \\ 0 & 0 & 0 & 2 & 0 \\ 0 & 0 & 0 & 0 & 2 \end{bmatrix}$$

The characteristic polynomial $(\lambda - 3)^3(\lambda - 2)^2$ reveals eigenvalues $\lambda_1 = 3$ with algebraic multiplicity 3 and $\lambda_2 = 2$ with algebraic multiplicity 2. Let us systematically construct the Jordan decomposition.

For $\lambda_1 = 3$, it is clear that $\ker(A - 3I) = \text{span}(1, 0, 0, 0, 0)^T$, showing geometric multiplicity 1. Two more vectors are required to complete this chain. Solving

$$(A - 3I)v_2 = \begin{pmatrix} 1 \\ 0 \\ 0 \\ 0 \\ 0 \end{pmatrix} \quad \Rightarrow \quad v_2 = \begin{pmatrix} a \\ 1/2 \\ 0 \\ 0 \\ 0 \end{pmatrix}$$

(where a is a free parameter) and then

$$(A - 3I)v_3 = \begin{pmatrix} a \\ 1/2 \\ 0 \\ 0 \\ 0 \end{pmatrix} \quad \Rightarrow \quad v_3 = \begin{pmatrix} b \\ a/2 + 1/4 \\ -1/2 \\ 0 \\ 0 \end{pmatrix}$$

(where b is another free parameter) provides our first chain $v_3 \mapsto v_2 \mapsto v_1 \mapsto \mathbf{0}$.

For $\lambda_2 = 2$, solving $(A - 2I)v = \mathbf{0}$ reveals $\lambda = 2$ to have geometric and algebraic multiplicity equal to 2, with independent eigenvectors

$$v_4 = \begin{pmatrix} 0 \\ 0 \\ 0 \\ 1 \\ 0 \end{pmatrix} \quad : \quad v_5 = \begin{pmatrix} -2 \\ 0 \\ 0 \\ 0 \\ 1 \end{pmatrix}$$

The transformation matrix V must assemble these chains as columns in order of decreasing chain length:

$$V = [v_3 \ v_2 \ v_1 \ v_4 \ v_5]$$

One checks that $a = 0$ and $b = 1/12$ suffice in this example to produce the Jordan form:

$$J = V^{-1}AV = \begin{bmatrix} 3 & 1 & 0 & 0 & 0 \\ 0 & 3 & 1 & 0 & 0 \\ 0 & 0 & 3 & 0 & 0 \\ 0 & 0 & 0 & 2 & 0 \\ 0 & 0 & 0 & 0 & 2 \end{bmatrix}$$

Ouch! The term "one checks" is doing the heavy lifting here...

The structure of J reflects both the algebraic multiplicities of eigenvalues and the lengths of generalized eigenvector chains discovered in our systematic decomposition.   ◇

Several practical principles emerge from this computation:

1. Work systematically through eigenvalues in order of decreasing algebraic multiplicity
2. For each eigenvalue, find all independent eigenvectors first
3. Build chains starting from eigenvectors not yet used in other chains
4. Assemble V by arranging chains in order of decreasing length
5. Verify the result through direct computation of AV and VJ

Two common pitfalls deserve special attention. First, when solving $(A - \lambda I)v_{i+1} = v_i$ for generalized eigenvectors, solutions always exist but contain free parameters. Different choices can produce different-looking but equivalent Jordan bases. Second, constructing V requires careful attention to column order – chains must be arranged to produce the correct block structure in J.

The practical computation of Jordan form thus unites the theoretical structures developed throughout this chapter. The filtration of generalized eigenspaces guides our search for chains, while careful basis construction transforms these chains into the similarity transformation V.

Though the process requires more subtlety than mere eigenvalue computation, systematic application of these principles yields this deepest decomposition of linear transformations.

8.5   Computing Eigenvalues

While focusing on the Jordan form and its computation, a preliminary step remains: how do we compute the eigenvalues in practice? The characteristic polynomial is not so simple – matrices of even modest size yield equations that resist direct solution. Eigenvalues of large matrices prompt more sophisticated methods of extraction. There is a remarkable algorithm based on a different matrix factorization that transforms our theoretical understanding into practical computation through careful composition of fundamental operations.

Consider first why naive approaches fail. A 100-by-100 matrix yields a degree-100 polynomial, far beyond the reach of standard root-finding methods. Even computing the coefficients proves perilous – expanding $\det(A - \lambda I)$ requires astronomical numbers of operations with catastrophic error growth. Yet modern software computes eigenvalues of thousand-by-thousand matrices in seconds. This seeming paradox resolves through iteration rather than direct solution.

The algorithm's elegance emerges from similarity transformations – a theme encountered repeatedly in our study of eigenvalues. If matrices A and B are similar, they share eigenvalues while potentially offering different computational advantages. The QR algorithm exploits this principle through careful iteration:

Definition 8.12 (QR Algorithm). Given matrix $A_0 = A$, the *QR algorithm* generates a sequence through:

1. Factor $A_k = Q_k R_k$ where Q_k is orthogonal and R_k is upper triangular
2. Form $A_{k+1} = R_k Q_k$ (reverse multiply)

Each iterate maintains similarity: $A_{k+1} = Q_k^T A_k Q_k$.   •

Though simple in description, this process harbors remarkable properties. The sequence $\{A_k\}$ converges to upper triangular form, with eigenvalues appearing on the diagonal. Complex conjugate pairs emerge naturally as 2-by-2 blocks, while repeated eigenvalues maintain their multiplicities. The algorithm effectively performs simultaneous power iteration on all eigenspaces, yet with crucial numerical stability through orthogonal transformations.

Example 8.13 (Simple Iteration). Consider the matrix

$$A_0 = \begin{bmatrix} 2 & 1 & 0 \\ 1 & 2 & 1 \\ 0 & 1 & 2 \end{bmatrix}$$

After five iterations we find:

$$A_5 \approx \begin{bmatrix} 3.414 & 0.892 & 0.247 \\ 0.000 & 2.000 & 0.731 \\ 0.000 & 0.000 & 0.586 \end{bmatrix}$$

The diagonal entries converge to eigenvalues approximately 3.414, 2.000, and 0.586 – roots of the characteristic polynomial $\lambda^3 - 6\lambda^2 + 11\lambda - 6$ that we never explicitly computed. $\diamond$

For matrices with complex eigenvalues, 2-by-2 blocks emerge naturally on the diagonal:

$$\begin{bmatrix} a & b \\ -b & a \end{bmatrix} \quad \text{representing} \quad \lambda = a \pm ib$$

This structure allows computation entirely in real arithmetic while still capturing complex behavior – a principle we first encountered when studying rotations in Section 8.1.

Modern implementations enhance this basic iteration through several refinements:

1. Shifts accelerate convergence by focusing iteration near likely eigenvalues

2. Implicit updates maintain structure while reducing operation count

3. Deflation techniques isolate converged eigenvalues for efficient processing

The algorithm's power emerges from its respect for matrix structure. Unlike characteristic polynomial computation, which discards geometry for pure algebra, QR iteration maintains crucial relationships:

- Orthogonal transformations preserve eigenvalues exactly
- Upper triangular form reveals eigenvalue approximations directly
- Similar matrices retain Jordan structure though it may be obscured

Though we lose explicit access to Jordan structure in the numerical process, the eigenvalues themselves emerge with remarkable accuracy. This balance – between theoretical understanding and practical computation – exemplifies how seemingly abstract concepts guide the development of efficient algorithms. The ideas that illuminated eigenvalue structure theoretically provide the tools for finding them in practice.

Example: The shifted QR algorithm subtracts an estimate μ of an eigenvalue, iterates, then adds μ back – dramatically improving convergence when μ is well-chosen.

8.6   Back to Basis

Our development of Jordan form illuminates the solution structure of higher-order linear differential equations. The abstract machinery of generalized eigenvectors and complex blocks translates directly into concrete solution patterns – polynomial terms multiplying exponentials, trigonometric functions emerging from complex pairs. This understanding completes our picture of basis solutions begun in Chapter 7, revealing why and how such patterns must arise.

Consider the general linear homogeneous differential equation of order n with degree n characteristic polynomial $p \in \mathcal{P}_n$:

$$p(D)x = (D^n + a_{n-1}D^{n-1} + \cdots + a_1 D + a_0)x = 0$$

When the roots of $p(\lambda) = 0$ are repeated, the companion matrix falls into Jordan blocks. The basis solutions emerge directly from the first row of the matrix exponential e^{Jt} computed in Section 8.3. For a Jordan block of size k with eigenvalue λ, we found:

$$e^{Jt} = e^{\lambda t}e^{Nt} = e^{\lambda t}\left(I + Nt + \frac{N^2 t^2}{2!} + \cdots + \frac{N^{k-1}t^{k-1}}{(k-1)!}\right)$$

The first row of this matrix yields our fundamental solutions:

Theorem 8.14 (Basis Solutions for Repeated Roots). *Let λ be a root of the characteristic equation $p(\lambda) = 0$ with algebraic multiplicity k. Then:*

1. *For real λ, the functions*

$$\phi_j(t) = t^j e^{\lambda t}, \quad j = 0, 1, \ldots, k-1$$

   form a basis for the solution space corresponding to λ.
2. *For complex $\lambda = \alpha \pm i\beta$, the functions*

$$\phi_j(t) = t^j e^{\alpha t}\cos(\beta t), \quad \psi_j(t) = t^j e^{\alpha t}\sin(\beta t), \quad j = 0, 1, \ldots, k-1$$

   form a real basis for the solution space corresponding to the conjugate pair.

These solutions arise directly from the first row of the exponentiated Jordan blocks computed in Section 8.3.

Example 8.15 (Repeated Real Root). The equation

$$\frac{d^3 x}{dt^3} - 3\frac{d^2 x}{dt^2} + 3\frac{dx}{dt} - x = 0$$

has characteristic polynomial $(\lambda - 1)^3 = 0$. The basis solutions are

$$\phi_0(t) = e^t, \quad \phi_1(t) = te^t, \quad \phi_2(t) = \frac{1}{2!}t^2 e^t$$

matching exactly the patterns seen in the Jordan form of its companion matrix. These emerge from the first row of the matrix exponential computed in our earlier example of a 3×3 Jordan block. The general solution is their linear combination:

$$x(t) = c_0 \phi_0(t) + c_1 \phi_1(t) + c_2 \phi_2(t)$$

$\diamond$

Example 8.16 (Damped Oscillator). The equation

$$\frac{d^2 x}{dt^2} + 2\gamma \frac{dx}{dt} + \omega^2 x = 0 \tag{8.5}$$

with $0 < \gamma < \omega$ (underdamped case) has characteristic equation

$$\lambda^2 + 2\gamma\lambda + \omega^2 = 0$$

with roots $\lambda = -\gamma \pm i\sqrt{\omega^2 - \gamma^2}$. The solution

$$x(t) = e^{-\gamma t}\left(c_1 \cos(\sqrt{\omega^2 - \gamma^2}\, t) + c_2 \sin(\sqrt{\omega^2 - \gamma^2}\, t)\right)$$

exhibits damped oscillations – exponential decay at rate γ modulating oscillations of frequency $\sqrt{\omega^2 - \gamma^2}$.

$\diamond$

Example 8.17 (Critical Damping). In equation (8.5), if we set the damping to the precise value $\gamma = \omega$, this represents critical damping when the damping coefficient exactly balances the natural frequency. Its characteristic polynomial

$$\lambda^2 + 2\omega\lambda + \omega^2 = (\lambda + \omega)^2$$

has repeated root $\lambda = -\omega$. The basis solutions are now

$$\phi_1(t) = e^{-\omega t} \quad \text{and} \quad \phi_2(t) = t e^{-\omega t}$$

Example: In mechanical systems, critical damping represents the fastest return to equilibrium without oscillation – all eigenvalues coincide at the negative point maximizing decay.

The transition between underdamped oscillation and pure decay manifests in this borderline case – the polynomial term t replacing trigonometric functions as the roots merge.

$\diamond$

Example 8.18 (Repeated Complex Roots). Consider the fourth-order equation

$$\frac{d^4 x}{dt^4} + 2\frac{d^2 x}{dt^2} + x = 0$$

whose characteristic polynomial $\lambda^4 + 2\lambda^2 + 1 = (\lambda^2 + 1)^2$ has root $\lambda = i$ with multiplicity 2. From the complex solutions $t^j e^{it}$, we obtain four real basis solutions:

$$\phi_1(t) = \cos(t), \quad \phi_2(t) = \sin(t), \quad \phi_3(t) = t\cos(t), \quad \phi_4(t) = t\sin(t)$$

These emerge directly from the first row of the exponentiated complex Jordan block, with the polynomial coefficients modulating pure sinusoidal motion to create more intricate oscillatory patterns. ◇

These examples illustrate the perfect correspondence between Jordan structure and scalar solutions. The nilpotent part of each Jordan block manifests as powers of t, while complex blocks generate the trigonometric functions essential to modeling oscillation. What appeared first as abstract matrix theory reveals itself as the natural language for describing physical motion.

The complete solution always emerges by combining the contributions from each Jordan block:

- Real eigenvalues yield exponential decay or growth
- Complex pairs generate oscillatory terms
- Repeated roots add polynomial factors

This decomposition – splitting motion into its fundamental modes – provides both theoretical insight and practical methods for understanding linear systems.

Power Grid Stability

The stability of electrical power grids provides a compelling demonstration of Jordan canonical form in practice. A modern power network consists of hundreds of generators linked by transmission lines, each machine contributing to a delicate dance of synchronized motion. When this dance falters, blackouts can cascade across continents. The mathematics of Chapter 8 reveals precisely how such instabilities develop through analysis of the network's Jordan structure.

Consider first a simple case: two generators connected by a transmission line. Each generator has an angle θ_i relative to the grid's nominal frequency. Small deviations from equilibrium follow linearized dynamics:

$$\begin{bmatrix} m_1 & 0 \\ 0 & m_2 \end{bmatrix} \begin{pmatrix} \ddot{\theta}_1 \\ \ddot{\theta}_2 \end{pmatrix} + \begin{bmatrix} d_1 & 0 \\ 0 & d_2 \end{bmatrix} \begin{pmatrix} \dot{\theta}_1 \\ \dot{\theta}_2 \end{pmatrix} + \begin{bmatrix} k & -k \\ -k & k \end{bmatrix} \begin{pmatrix} \theta_1 \\ \theta_2 \end{pmatrix} = 0$$

where m_i represents generator inertia, d_i damping, and k the transmission line coupling. Converting to first-order form by introducing phase angles and frequencies as state variables yields a 4×4 system matrix A.

The crucial feature appears when generators have identical parameters: $m_1 = m_2 = m$ and $d_1 = d_2 = d$. Then A develops eigenvalues of algebraic multiplicity two:

$$A = \begin{bmatrix} 0 & 0 & 1 & 0 \\ 0 & 0 & 0 & 1 \\ -k/m & k/m & -d/m & 0 \\ k/m & -k/m & 0 & -d/m \end{bmatrix}$$

Historical Note: The 2003 Northeast blackout, affecting 50 million people, began when overloaded transmission lines sagged into trees. The resulting loss of synchronization demonstrated dramatically how local instabilities can propagate through grid structures.

Direct computation shows these eigenvalues come in pairs:

$$\lambda_{1,2} = -\frac{d}{2m} \pm i\sqrt{\frac{2k}{m} - \frac{d^2}{4m^2}}$$

each with algebraic multiplicity 2 but geometric multiplicity 1. The physical meaning is clear: identical generators can oscillate in phase or anti-phase, with complex eigenvalues indicating oscillatory motion.

Following Section 8.3, this matrix has Jordan canonical form:

$$A = PJP^{-1}$$

where J takes real Jordan form with two 2×2 blocks:

$$J = \begin{bmatrix} C & I_2 \\ 0 & C \end{bmatrix}, \quad C = \begin{bmatrix} -d/(2m) & -\omega \\ \omega & -d/(2m) \end{bmatrix}$$

where $\omega = \sqrt{2k/m - d^2/(4m^2)}$ gives the oscillation frequency. This real form, as developed in Section 8.3, reveals both the damped oscillation through C and the coupling between modes through I_2.

The appearance of nontrivial Jordan structure signals subtle instability: though the system appears symmetric, small perturbations can excite growing oscillations as generators lose synchronization. Grid operators must carefully monitor these near-resonant conditions where eigenvalues coalesce.

Real power networks involve hundreds of generators, leading to stability matrices of enormous dimension. Computing their Jordan structure demands the careful numerical methods developed in Section 8.5. The QR algorithms prove essential, as direct eigenvalue computation becomes unstable precisely when Jordan blocks begin to form — exactly the dangerous regime operators must monitor.

Modern grid analysis employs sophisticated variants of these methods. Rather than compute full Jordan decompositions (which prove numerically delicate), operators track the migration of eigenvalues through the complex plane using the robust iteration schemes of Section 12.3. When eigenvalues approach coalescence, indicating incipient Jordan block formation, the network can be reconfigured to maintain stability.

The key insight is that Jordan structure in power networks indicates more than mere mathematical curiosity — it signals physical conditions requiring immediate attention. Each identity superdiagonal in a Jordan block represents a path for instability to propagate between generators. Understanding this structure proves crucial for preventing cascading failures.

Grid stability thus exemplifies how the mathematics of Jordan canonical form shapes critical infrastructure. The delicate interplay between complex eigenvalues and geometric multiplicity, seemingly abstract in Section 8.2, manifests physically in the synchronized dance of massive generators. Modern grid reliability depends fundamentally on the careful numerical computation of this structure.

The profound lesson is that apparently identical components — like generators with matching parameters — create subtle dangers through their identical eigenvalues. The Jordan canonical form reveals precisely how such symmetry

Example: Modern grid control systems continuously assess stability margins using a combination of model-based analysis and real-time measurements. When these assessments, which are related to the system's eigenvalues, indicate a reduction in stability margins or a risk of growing oscillations, operators receive alerts to adjust system conditions.

Nota bene: Unlike mechanical resonance, power grid instability through Jordan block formation can occur without external forcing — a sobering reminder of how mathematical structure alone can precipitate physical disaster.

enables instability to propagate. Every major power grid control center now employs these mathematical tools, developed in this chapter, to maintain the reliable electricity supply on which modern society depends.

Chemical Process Network Analysis

The identification of weakly coupled subsystems within large chemical processing networks demonstrates another powerful application of Jordan canonical form analysis. Modern chemical plants often combine dozens of reaction vessels, heat exchangers, and separation units into complex networks where material and energy flows create subtle couplings between processes. Understanding these couplings – particularly which processes can be treated as nearly independent – proves crucial for both safety and efficient control.

Consider a chemical plant with six coupled reaction vessels. In each vessel i, both the concentration $c_i(t)$ of a key reactant and the temperature $T_i(t)$ evolve according to coupled differential equations. Material flows between vessels create concentration coupling, while shared cooling systems introduce thermal coupling. Let $x = (c_1, T_1, \ldots, c_6, T_6)^T$ represent the complete state. Under standard assumptions of mass action kinetics and Newtonian cooling, the system takes the form:

$$\frac{dx}{dt} = Ax$$

For a typical petrochemical process operating near steady state, we might find coupling matrix:

$$A = \begin{bmatrix} A_{11} & A_{12} & \varepsilon_{13} \\ A_{21} & A_{22} & 0 \\ \varepsilon_{31} & 0 & A_{33} \end{bmatrix}$$

where each A_{ii} is a 4×4 block describing two coupled vessels, A_{ij} represents stronger coupling between adjacent units, and ε_{ij} indicates weak coupling through shared utilities. Taking realistic values for an ethylene oxide production unit:

$$A_{11} = \begin{bmatrix} -2.1 & 0.8 & 0.5 & 0.1 \\ 0.6 & -1.7 & 0.1 & 0.4 \\ 0.5 & 0.1 & -1.9 & 0.7 \\ 0.1 & 0.4 & 0.5 & -1.6 \end{bmatrix}$$

with similar structure for A_{22} and A_{33}, while the coupling terms have entries of order 0.1 or 0.01 for strong and weak coupling respectively.

Computing eigenvalues through the methods of Section 8.5 reveals three distinct clusters, each parameterized by the coupling strength ε:

$$\lambda_1 = -2.2 + O(\varepsilon), \quad \lambda_2 = -2.2 + O(\varepsilon)$$

$$\lambda_3, \lambda_4, \lambda_5 = -1.8 + O(\varepsilon)$$

$$\lambda_6, \ldots, \lambda_{12} = -1.5 + O(\varepsilon)$$

Example: In ethylene oxide production, reactor temperatures typically range from 200-300°C, making thermal coupling through shared cooling systems a significant safety consideration.

These clustered eigenvalues suggest near-independence between process units. The Jordan decomposition confirms this structure:

$$J = \begin{bmatrix} J_1 & \varepsilon E_{12} & \varepsilon E_{13} \\ \varepsilon E_{21} & J_2 & \varepsilon E_{23} \\ \varepsilon E_{31} & \varepsilon E_{32} & J_3 \end{bmatrix}$$

where each J_i corresponds to a cluster of eigenvalues and each E_{ij} is a matrix with entries of order unity, so the off-diagonal blocks contain terms of order ε.

This decomposition has profound implications for process control and safety:

1. Each process train can be controlled quasi-independently
2. Disturbances remain largely confined within blocks
3. Critical failure modes align with Jordan block structure
4. Sensor placement should respect subsystem boundaries

The generalized eigenvectors associated with each Jordan block reveal exactly how disturbances propagate between units. For example, a temperature excursion in vessel 1 affects its paired vessel 2 strongly but influences the other process trains only weakly through the $O(\varepsilon)$ terms. This understanding proves crucial for safety system design.

Historical Note: Major chemical accidents like the 1974 Flixborough disaster have highlighted the importance of understanding how process disturbances can propagate through weakly coupled systems.

Modern chemical plant design explicitly considers this block structure. Critical processes are grouped to minimize coupling with other units, while safety systems are designed around identified subsystem boundaries. The Jordan canonical form provides the mathematical framework for quantifying these design decisions.

The computational methods of Section 8.5 prove essential for large plants, where direct eigenvalue calculation becomes numerically unstable precisely when identifying weak coupling is most important. The QR iteration reliably reveals clustered eigenvalues even when direct methods fail.

This chemical engineering application complements our flutter analysis by showing how Jordan structure illuminates weakly coupled systems. Where aircraft exhibit exact repeated eigenvalues from physical symmetries, chemical processes display near-repetition from weak coupling. Together they demonstrate how the mathematical framework developed in this chapter guides both analysis and design of complex engineered systems.

Exercises: Chapter 8

1. The matrix

$$A = \begin{bmatrix} 5 & -12 \\ 4 & 5 \end{bmatrix}$$

has complex eigenvalues. Find them and write e^{At} in terms of trigonometric functions. Describe geometrically what this matrix does to vectors in $\mathbb{R}^2$ as t increases.

2. Find the eigenvalues and generalized eigenvectors of

$$A = \begin{bmatrix} 3 & 1 & 0 & 0 \\ 0 & 3 & 1 & 0 \\ 0 & 0 & 3 & 1 \\ 0 & 0 & 0 & 3 \end{bmatrix}$$

Then use them to compute e^{At}. Verify your answer by checking that it satisfies both $(d/dt)e^{At} = Ae^{At}$ and $e^{A0} = I$.

3. Consider the 5×5 matrix

$$A = \begin{bmatrix} 3 & 1 & 0 & 0 & 0 \\ 0 & 3 & 0 & 0 & 0 \\ 0 & 0 & -1 & 1 & 0 \\ 0 & 0 & 0 & -1 & 1 \\ 0 & 0 & 0 & 0 & -1 \end{bmatrix}$$

Find its Jordan canonical form and compute e^{At}. Which solutions grow as $t \to \infty$?

4. For the system

$$\frac{d}{dt}\begin{pmatrix} x \\ y \\ z \\ w \end{pmatrix} = \begin{bmatrix} -1 & 1 & 0 & 0 \\ 0 & -1 & 0 & 0 \\ 0 & 0 & -1 & 1 \\ 0 & 0 & 0 & -1 \end{bmatrix}\begin{pmatrix} x \\ y \\ z \\ w \end{pmatrix}$$

find all solutions that remain bounded as $t \to \infty$. Show that these form a subspace and find its dimension.

5. Consider the matrix

$$A = \begin{bmatrix} -1 & -2 & 1 & 0 & 0 \\ 2 & -1 & 0 & 1 & 0 \\ 0 & 0 & 4 & 1 & 0 \\ 0 & 0 & 0 & 4 & 1 \\ 0 & 0 & 0 & 0 & 4 \end{bmatrix}$$

Find its Jordan canonical form and a matrix V such that $V^{-1}AV$ gives this form. What does e^{At} look like?

6. For the fourth-order equation

$$\frac{d^4x}{dt^4} + 4\frac{d^2x}{dt^2} + 4x = 0$$

find all linearly independent solutions. (Hint: the characteristic polynomial factors as $(\lambda^2 + 2)^2$.)

7. Consider the matrices

$$A = \begin{bmatrix} 1 & 1 & 0 & 0 & 0 & 0 \\ 0 & 1 & 0 & 0 & 0 & 0 \\ 0 & 0 & 1 & 1 & 0 & 0 \\ 0 & 0 & 0 & 1 & 1 & 0 \\ 0 & 0 & 0 & 0 & 1 & 0 \\ 0 & 0 & 0 & 0 & 0 & 1 \end{bmatrix} \quad \text{and} \quad B = \begin{bmatrix} 1 & 1 & 0 & 0 & 0 & 0 \\ 0 & 1 & 1 & 0 & 0 & 0 \\ 0 & 0 & 1 & 0 & 0 & 0 \\ 0 & 0 & 0 & 1 & 1 & 0 \\ 0 & 0 & 0 & -1 & 1 & 0 \\ 0 & 0 & 0 & 0 & 0 & 1 \end{bmatrix}$$

For each matrix: list all eigenvalues with their algebraic and geometric multiplicities and identify the Jordan blocks and their sizes; then, compute e^{At} and e^{Bt}.

8. Let A be a 4×4 matrix with characteristic polynomial $p(\lambda) = (\lambda - 5)^4$. How many different Jordan canonical forms are possible for A? Draw diagrams showing all possibilities.

9. For a real 4×4 matrix A with complex conjugate pairs of eigenvalues $1 \pm 2i$ and $-3 \pm i$, what are the possible Jordan canonical forms? Draw diagrams showing all possibilities.

10. Let

$$
A = \begin{bmatrix} 1 & -4 & 2 & 0 \\ 1 & 1 & 0 & 2 \\ 0 & 0 & 3 & 1 \\ 0 & 0 & 0 & 3 \end{bmatrix}
$$

Find all possible Jordan canonical forms consistent with its eigenvalues. Which is correct? Explain how to determine this without computing V.

11. Consider the matrix

$$
A = \begin{bmatrix} 3 & -4 & 1 & 0 & 0 \\ 4 & 3 & 0 & 1 & 0 \\ 0 & 0 & -2 & 1 & 0 \\ 0 & 0 & 0 & -2 & 1 \\ 0 & 0 & 0 & 0 & -2 \end{bmatrix}
$$

Find its Jordan canonical form. What is the minimal polynomial of A?

12. Show that for any nilpotent matrix N of size $n \times n$, the series

$$
I + tN + \frac{t^2}{2!}N^2 + \frac{t^3}{3!}N^3 + \cdots + \frac{t^{n-1}}{(n-1)!}N^{n-1}
$$

gives e^{Nt} exactly. Use this to find e^{Nt} for the 4×4 nilpotent matrix N with ones on the superdiagonal.

13. Let A be a real $n \times n$ matrix. Prove that if λ is an eigenvalue of A with algebraic multiplicity m, then the solutions to $(A - \lambda I)^m x = 0$ form a subspace of dimension exactly m.

14. For a 3×3 matrix A with repeated eigenvalue λ, prove that there are exactly three possible Jordan canonical forms. Draw diagrams showing the generalized eigenvector chains in each case.

15. A 4×4 matrix A has characteristic polynomial $(\lambda + 1)^2(\lambda - 3)(\lambda + 2)$. If $\ker(A + I)$ is one-dimensional, what must the Jordan canonical form of A be? Justify your answer.

16. For an $n \times n$ matrix A with eigenvalues $\lambda_1, \ldots, \lambda_n$ (counted with multiplicity), prove that $\operatorname{tr}(e^{At}) = \sum_{k=1}^{n} e^{\lambda_k t}$. Use this to show that if all eigenvalues have negative real part, then $\operatorname{tr}(e^{At}) \to 0$ as $t \to \infty$.

17. Let A be a real 2×2 matrix with complex eigenvalues $a \pm bi$ where $b \neq 0$. Prove that A cannot be written as a product of two real matrices with real eigenvalues. (Hint: consider what happens to eigenvalues under matrix multiplication).

18. Let A be a real 4×4 matrix with eigenvalues $3 \pm 4i$ and $-2 \pm i$. What form must

$$\lim_{t \to \infty} \frac{e^{At}}{e^{3t}}$$

   take? Justify your answer.

19. Show that for any $n \times n$ matrix A, the nilpotent part N in its Jordan decomposition satisfies $N^n = 0$. Give an example of a 5×5 nilpotent matrix N where $N^4 \neq 0$ but $N^5 = 0$.

20. Use Cayley-Hamilton to prove that if A is nilpotent (meaning some power equals zero) then its characteristic polynomial must be $p(\lambda) = \lambda^n$ where n is the size of the matrix. What does this tell you about the eigenvalues of nilpotent matrices?

21. Let A be an $n \times n$ matrix. Prove that the minimal polynomial of A (the monic polynomial of least degree that annihilates A) must divide the characteristic polynomial. How does this relate to the Jordan canonical form?

See the Chapter 7 exercises for the Cayley-Hamilton Theorem.

Chapter 9
Linear Iterative Systems

POWER BEGETS POWER in the iteration of linear transformations. Each application of a matrix to a vector shifts weight toward dominant directions, channeling through preferred pathways until some natural balance emerges. This fundamental process – the convergence of repeated transformation toward dominant modes – pervades both theory and computation. Though simpler than the continuous flows and Jordan structures of previous chapters, these iterative systems harbor their own subtle beauty in the delicate interplay between matrix structure and asymptotic behavior.

The central idea lies in the connection between matrix powers and eigenvalues. In the cleanest case, iterating a linear transformation leads to one eigenvalue dominating the results, dragging all vectors toward its eigendirection. This process distills a matrix to its essential axis. The power method for finding largest eigenvalues emerges naturally from the iteration itself, transforming a theoretical insight into a computational tool.

Yet not all matrices yield so readily to analysis. Special structures – positivity constraints, probability conservation, symmetry – all complicate and enlighten our study. The Perron-Frobenius theory reveals how positivity forces uniqueness of dominant directions. Stochastic matrices preserve total measure while driving systems toward equilibrium states. Graph matrices encode discrete topology through their spectrum. Each class brings its own spectral features that shape long-term behavior.

Our development moves from the abstract to the concrete, from general iteration to structured classes of matrices whose special properties illuminate both theory and application. The patterns we uncover extend naturally from finite to infinite dimensions, from discrete steps to continuous limits. Throughout, we maintain the perspective that iteration provides both theoretical insight and practical power – each application of a transformation bringing us closer to understanding its fundamental nature.

9.1   At First Iteration

The Fibonacci sequence provides a first glimpse of the mysteries inherent in linear iteration. Each number, the sum of its two predecessors, generates the sequence

$$0, 1, 1, 2, 3, 5, 8, 13, 21, 34, 55, 89, 144, \ldots$$

More striking than the numbers themselves is a hidden pattern: the ratio of consecutive terms approaches a fixed value $\varphi = (1 + \sqrt{5})/2 \approx 1.618$. This convergence of ratios hints at deeper structure within linear recurrence relations.

Yes, this is the so-called *golden ratio*. No, it is not here because it is ubiquitous in art & nature. It appears here because it is a dominant eigenvalue: keep reading.

The Fibonacci rule $x_{n+2} = x_{n+1} + x_n$ represents a second-order recurrence. Just as we transformed second-order differential equations to first-order systems in Chapter 7, we can recast this scalar sequence as a vector iteration. Setting

$$v(n) = \begin{pmatrix} x_n \\ x_{n+1} \end{pmatrix}$$

transforms the recurrence into

$$v(n + 1) = \begin{bmatrix} 0 & 1 \\ 1 & 1 \end{bmatrix} v(n)$$

This first-order vector system captures the same evolution through matrix multiplication. Its solution is clear if not clearly computable:

$$v(n) = A^n v(0) = A^n \begin{pmatrix} 1 \\ 1 \end{pmatrix}$$

More generally, a discrete-time linear system evolving has form and solution

$$x(n + 1) = Ax(n) \quad \Rightarrow \quad x(n) = A^n x(0) \tag{9.1}$$

where A encodes the rules of evolution. Just as differential equations $dx/dt = Ax$ generate continuous flow through infinitesimal change, recurrence relations (9.1) create discrete trajectories through stepped iteration.

This discrete framework finds natural application in economic systems, where regular time periods (quarters, years) impose inherent granularity.

Consider an economy divided into n sectors, each producing goods consumed partly by other sectors and partly as final output. The *input-output matrix* $A = [a_{ij}]$ encodes production requirements: entry a_{ij} represents the amount of sector i's output needed to produce one unit of sector j's output. The evolution of such a system follows equation (9.1), where $x(n)$ represents sector outputs in period n.

Example 9.1 (Six-Sector Economy). Consider an economy with six primary sectors:

1. Agriculture (food production)
2. Energy (power generation)
3. Manufacturing (industrial goods)
4. Transportation (logistics)
5. Services (business/consumer)
6. Technology (IT/communications)

The input-output matrix A captures their interdependencies:

$$A = \begin{bmatrix} 0.15 & 0.08 & 0.10 & 0.05 & 0.20 & 0.05 \\ 0.25 & 0.30 & 0.35 & 0.40 & 0.20 & 0.30 \\ 0.10 & 0.15 & 0.20 & 0.25 & 0.10 & 0.20 \\ 0.15 & 0.12 & 0.15 & 0.15 & 0.10 & 0.08 \\ 0.20 & 0.25 & 0.15 & 0.15 & 0.25 & 0.30 \\ 0.15 & 0.15 & 0.20 & 0.15 & 0.25 & 0.20 \end{bmatrix}$$

Each column represents input requirements for one unit of sector output. For instance, producing one unit of manufacturing output (column 3) requires 0.10 units of agricultural input, 0.35 units of energy, and so forth. The columns sum to about 0.95 as they exclude labor and other external factors.

A remarkable pattern emerges under iteration. Regardless of initial conditions, the relative sizes of sectors converge to fixed proportions approximately equal to $(0.22, 0.62, 0.34, 0.25, 0.48, 0.39)^T$, while the overall economy grows by roughly 1.09 (or 9%) each period. This means that after sufficient time, the energy sector (2) stabilizes at almost 2.8 times the size of the agricultural sector (1), the service sector (5) maintains about 1.2 times the size of the technology sector (6), and so forth – despite widely varying initial conditions and complex interconnections. $\diamond$

   This tension – between the apparent complexity of sector-wise interaction and the simplicity of asymptotic behavior – exemplifies a broader principle in linear systems. Complex coupling through multiple variables often reduces to simpler patterns driven by dominant modes. Our task in

Historical Note: This type of economic model is called a *Leontief input-output model*, after W. Leontief who developed it in the 1930s through detailed study of the American economy. His work on structural interdependence in production earned him the 1973 Nobel Prize in Economics. Though originally developed using data laboriously collected by hand, such models now inform economic planning worldwide through automated data collection and computation.

Foreshadowing: The uniform growth rate emerging from Leontief models previews the Perron-Frobenius theory of positive matrices, where special structure ensures uniqueness of a dominant positive eigenvalue.

this chapter is to understand this reduction, revealing how eigenstructure shapes the long-term character of linear iteration.

9.2   Dominance & Convergence

The mysterious convergence of Fibonacci ratios hints at deeper structure within matrix powers. Just as the ratio sequence x_{n+1}/x_n approaches the golden mean φ, general matrix iterations often display similar convergence – their behavior dominated by a single eigenvalue that drives long-term evolution. This reduction of complex iteration to simple scaling reflects a fundamental principle: repeated linear transformation distills a matrix to its essential character.

The key to understanding such convergence lies in the representation of matrix powers from Lemma 7.7. For a diagonalizable matrix $A = V\Lambda V^{-1}$, powers take the form

$$A^n = V\Lambda^n V^{-1}$$

where Λ^n simply raises each eigenvalue to the nth power. When these eigenvalues differ in magnitude, larger ones grow faster than smaller ones, eventually dominating the iteration. This observation leads to crucial definitions:

Definition 9.2 (Spectral Radius and Dominance). The *spectral radius* ρ_A of a matrix A is the maximum magnitude of its eigenvalues:

$$\rho_A = \max\{|\lambda| : \lambda \text{ is an eigenvalue of } A\}$$

An eigenvalue λ_* is *dominant* if $|\lambda_*| = \rho_A$ and no other eigenvalue has this magnitude. Its corresponding eigenvector v_* is called the *dominant eigenvector*.

Lemma 9.3 (Dominant Convergence). *Let A be diagonalizable with dominant eigenvalue λ_* and corresponding eigenvector v_*. Then for any initial vector x_0 with nonzero component C along v_*,*

$$A^n x_0 \simeq C\lambda_*^n v_* \text{ as } n \to \infty$$

Proof. Index the eigenvalues $\{\lambda_i\}$ such that $\lambda_* = \lambda_1$. Writing x_0 in the eigenbasis $\{v_i\}$ of A and applying A^n yields

$$A^n x_0 = A^n \sum_{k=1}^{n} c_k v_k = \sum_{k=1}^{n} c_k \lambda_k^n v_k \tag{9.2}$$

$$= c_1 \lambda_*^n \left(v_* + \sum_{k=2}^{n} \frac{c_k}{c_1} \left(\frac{\lambda_k}{\lambda_*} \right)^n v_k \right) \simeq c_1 \lambda_*^n v_* \tag{9.3}$$

Nota bene: The notion of eigenvalue dominance holds sway in ODEs and continuous-time dynamics as well; however, it is not the spectral radius that matters for continuous-time dominance – it is the real parts of eigenvalues that count. Why? The differentiation operator D is the natural logarithm of the shift operator E.

Since $|\lambda_k/\lambda_1| < 1$ for all $k > 1$, these ratios decay exponentially to zero as $n \to \infty$. The constant $c_1 = C$. $\qquad\square$

This convergence principle underlies the *power method* for computing dominant eigenvalues. Starting with a random vector x_0, repeated multiplication by A followed by normalization yields convergence to the dominant eigendirection. The rate of convergence is governed by the ratio of the second largest eigenvalue magnitude to the dominant one.

Example 9.4 (Fibonacci redux). For the Fibonacci matrix of the previous section,

$$A = \begin{bmatrix} 0 & 1 \\ 1 & 1 \end{bmatrix}$$

the eigenvalues are $\varphi = (1 + \sqrt{5})/2$ and $\psi = (1 - \sqrt{5})/2$, with $|\varphi| > 1 > |\psi|$. As $n \to \infty$, the powers of the dominant eigenvalue φ^n grow exponentially faster than $|\psi|^n$, explaining why consecutive Fibonacci numbers approach ratio φ. $\qquad\diamond$

Example 9.5 (Economic Convergence). Returning to Example 9.1, the six-sector economy's input-output matrix A has dominant eigenvalue $\lambda_* = \rho_A \approx 1.09$ with corresponding dominant eigenvector

$$v_* \approx (0.22, 0.62, 0.34, 0.25, 0.48, 0.39)^T$$

normalized to unit length. This explains both the 9% growth rate and the fixed proportions observed empirically – the economy's structure forces convergence to these ratios regardless of initial conditions. $\qquad\diamond$

When A is not diagonalizable, Jordan blocks complicate but do not fundamentally alter this picture. Powers of a Jordan block

$$J_k = \lambda I + N$$

involve binomial terms:

$$J_k^n = \lambda^n \left(I + \frac{n}{\lambda} N + \frac{n(n-1)}{2\lambda^2} N^2 + \cdots \right)$$

The polynomial factors in n modify but cannot overcome the exponential dominance of the largest $|\lambda|$. Thus spectral radius still controls asymptotic behavior, though convergence may display subtle oscillations from the Jordan structure.

This reduction of matrix iteration to dominant modes exemplifies a broader principle: complex linear systems often simplify dramatically under repeated action. Whether modeling economic growth, population dynamics, or network influence, the long-term behavior typically reflects just one or two characteristic directions. The art lies in recognizing when such reduction occurs and exploiting the resulting simplification.

9.3   *Positivity & Perron-Frobenius Theory*

Certain natural quantities like mass, energy, and population are bound below by zero; probabilities and measures alike stubbornly refuse to venture into negative territory. When linear transformations act on such intrinsically positive quantities, their matrices inherit special structure that profoundly shapes their spectral properties. This structure – the intrinsic positivity of real-world measurements – leads to remarkable conclusions about eigenvalues and asymptotic behavior.

Consider a matrix $A = [a_{ij}]$ whose entries are all positive: $a_{ij} > 0$ for all i, j. Such matrices arise naturally in modeling coupled growth processes, where each component positively influences all others. The Perron-Frobenius theory reveals that such matrices possess uniquely simple spectral properties:

Theorem 9.6 (Perron-Frobenius). *Let A be a square matrix with all entries strictly positive. Then A has a dominant eigenvalue $\lambda_* = \rho_A > 0$, with corresponding dominant eigenvector $v_* > 0$ having all positive components. Moreover, any nonnegative eigenvector of A must lie in the dominant eigenspace.*

The proof of this remarkable theorem illuminates how positivity shapes spectral structure. Consider iteration of A on any positive vector x_0. The sequence of normalized vectors $y_n = A^n x_0 / \|A^n x_0\|$ remains positive, and a key inequality emerges:

$$\min_i \frac{(Ay)_i}{(y)_i} \leq \rho_A \leq \max_i \frac{(Ay)_i}{(y)_i}$$

holding for any positive vector y. These bounds, forced by positivity, trap the spectral radius. A delicate argument shows the bounds converge as $n \to \infty$, yielding both the dominant eigenvalue and its positive eigenvector. The impossibility of maintaining positivity through oscillation or decay then forces all other eigenvalues to have strictly smaller magnitude.

Example 9.7 (Research Citation Network). Consider the influence network among six major research areas in computer science, where entry b_{ij} of a matrix B represents the relative citation flow: of all citations *received* by papers in field j during a given year, what fraction came *from* papers in field i. The field-level averaging reveals citation patterns and a Perron-Frobenius dominant eigenvector as follows:

$$B = \begin{bmatrix} 0.60 & 0.25 & 0.15 & 0.10 & 0.05 & 0.05 \\ 0.20 & 0.50 & 0.15 & 0.10 & 0.05 & 0.05 \\ 0.10 & 0.10 & 0.50 & 0.15 & 0.05 & 0.05 \\ 0.05 & 0.05 & 0.10 & 0.50 & 0.10 & 0.05 \\ 0.03 & 0.05 & 0.05 & 0.10 & 0.65 & 0.10 \\ 0.02 & 0.05 & 0.05 & 0.05 & 0.10 & 0.70 \end{bmatrix} \quad : \quad v_* \approx \begin{pmatrix} 0.220 \\ 0.250 \\ 0.180 \\ 0.148 \\ 0.122 \\ 0.081 \end{pmatrix}$$

These correspond to: (1) Machine Learning, (2) Artificial Intelligence, (3) Computer Vision, (4) Robotics, (5) Systems & Networks, and (6) Theory. For instance, when examining all citations received by AI papers (column 2), 25% ($b_{12} = 0.25$) come from ML papers, 50% ($b_{22} = 0.50$) come from other AI papers, and smaller fractions come from other fields. The Perron-Frobenius eigenvector v_* (the eigenvector corresponding to $\lambda_* = 1$) provides a natural ranking of field influence or the stationary distribution of citations, with AI and ML dominating but maintaining significant coupling to application areas. The dominant eigenvalue is $\lambda_* = 1$; the next largest eigenvalue magnitude is $|\lambda_2| \approx 0.647$. This reveals the convergence rate to the stationary distribution v_*; the influence of other eigenmodes decays by a factor of ≈ 0.647 with each citation cycle. Convergence to within a few percent of the final distribution occurs relatively quickly, roughly within N cycles where 0.647^N is small (e.g., $N \approx 5$ cycles for decay to 10%). $\diamond$

The power of these results extends beyond strictly positive matrices. A matrix A is called *nonnegative* if all its entries are ≥ 0, and *irreducible* if no coordinate permutation can transform it into block triangular form. These properties capture the connectivity of the underlying interaction network: irreducibility means every component eventually influences every other component, either directly or through intermediaries.

Theorem 9.8 (Frobenius Extension). *The conclusions of the Perron-Frobenius theorem hold for nonnegative irreducible matrices, with the spectral radius $\rho_A > 0$ still positive but possibly having complex conjugate eigenvalues of magnitude ρ_A.*

Example 9.9 (Catalytic Cycle). Consider a chemical reaction network where substrates S_1, S_2, S_3, S_4 convert cyclically with transition matrix

$$
\begin{array}{cc}
\begin{array}{c}
S_1 \xRightarrow{\;k_1\;} S_2 \\
k_4 \Big\Uparrow \qquad \Big\Downarrow k_2 \\
S_4 \xLeftarrow[k_3]{} S_3
\end{array}
& : \quad A = \begin{bmatrix} 0 & 0 & 0 & k_4 \\ k_1 & 0 & 0 & 0 \\ 0 & k_2 & 0 & 0 \\ 0 & 0 & k_3 & 0 \end{bmatrix}
\end{array}
$$

with each reaction catalyzed by an enzyme. The characteristic polynomial of A is $\lambda^4 - k_1 k_2 k_3 k_4 = 0$ and the eigenvalues lie equally-spaced on the circle in the complex plane of radius $\rho_A = (k_1 k_2 k_3 k_4)^{1/4}$. The cyclic nature of the reaction is reflected in these two complex pairs of eigenvalues, while irreducibility ensures the existence of a positive steady-state distribution among the substrates. $\qquad\qquad\qquad\qquad\qquad\qquad\diamond$

> Biochemists call such networks *futile cycles* when they appear to serve no purpose beyond energy consumption, though they often play crucial regulatory roles in cellular metabolism.

The connection between positivity and spectral properties exemplifies a broader principle: constraints on matrix entries often force constraints on eigenstructure. Understanding these relationships – how matrix structure shapes spectral behavior – provides both theoretical insight and practical tools for analyzing linear systems. The Perron-Frobenius theory stands as perhaps the most elegant example of this principle, where simple positivity yields profound conclusions about asymptotic behavior.

9.4 Stochastic Matrices & Markov Chains

Many real-world processes involve transitions between states where probabilities govern the changes. Examples include weather patterns shifting between sunny and rainy conditions, animal populations moving between territories, and genetic traits passing between generations. Such processes, where future states depend only on the present and not on past history, lead naturally to the theory of Markov chains – a framework unifying discrete random processes through linear algebra.

Definition 9.10 (Markov Chain). A *Markov chain* is a sequence of random variables $\{X_n\}_{n\geq 0}$ taking values in a set of states S, satisfying the *Markov property*:

$$
\mathbb{P}(X_{n+1} = j \mid X_n = i, X_{n-1} = i_{n-1}, \ldots, X_0 = i_0) = \mathbb{P}(X_{n+1} = j \mid X_n = i)
$$

> *Relax...* If you have not seen conditional probability, you can proceed without worrying too much about it. When you get the chance, read up on conditionals. For now, just work with P as a matrix of probabilities.

The probability p_{ij} of transitioning from state i to state j in one step defines an entry in the chain's *transition matrix* $P = [p_{ij}]$. $\qquad\bullet$

The Markov property – that future depends on present but not past – transforms temporal evolution into matrix iteration. To see this connection, we must first clarify what our vectors represent:

Definition 9.11 (Probability Distribution). A vector $x = (x_1, \ldots, x_n)^T$ is a *probability distribution* if:

1. Nonnegativity: $x_i \geq 0$ for all i
2. Total probability: $\sum_{i=1}^{n} x_i = 1$

The entry x_i represents the probability of being in state i. •

The transition matrices of Markov chains must preserve these probability constraints, leading to our central object of study:

Definition 9.12 (Stochastic Matrix). A square matrix $P = [p_{ij}]$ is *stochastic* if it satisfies:

1. Nonnegativity: $p_{ij} \geq 0$ for all i, j
2. Column-stochasticity: $\sum_{i=1}^{n} p_{ij} = 1$ for all j

The entry p_{ij} represents the probability of transitioning from state j to state i in one step. •

Caveat: many authors use row-stochastic matrices, where the rows are probability distributions. If the term *stochastic* is used, always double-check which type is meant.

The evolution of probabilities in a Markov chain follows our familiar iteration pattern:

$$x(k+1) = Px(k)$$

where $x(k)$ represents the probability distribution across states at step k. The stochastic constraints on P ensure that if $x(k)$ is a probability distribution, then $x(k+1)$ will be as well.

Example 9.13 (Weather Patterns). Consider a simple model of daily weather transitions among three states: Sunny (S), Cloudy (C), and Rainy (R). Historical data suggests transition probabilities:

$$P = \begin{bmatrix} 0.7 & 0.3 & 0.2 \\ 0.2 & 0.4 & 0.3 \\ 0.1 & 0.3 & 0.5 \end{bmatrix}$$

Reading down columns: from a sunny day (first column), the weather transitions to sunny with probability 0.7, cloudy with 0.2, and rainy with 0.1; similarly for transitions from cloudy or rainy states.

Given initial distribution $x(0) = (1, 0, 0)^T$ representing certainty of sun today, tomorrow's distribution becomes:

$$x(1) = Px(0) = (0.7, 0.2, 0.1)^T$$

showing how probability disperses across states. After two days:

$$x(2) = P^2 x(0) = (0.55, 0.27, 0.18)^T$$

suggesting convergence toward some equilibrium distribution. ◇

The spectral properties of stochastic matrices determine their long-term behavior. A Markov chain (and its transition matrix P) is *irreducible* if it is possible to reach any state from any other state (not necessarily in one step). It is *aperiodic* if the greatest common divisor of the lengths of all possible return paths to any state is 1. An irreducible and aperiodic Markov chain is called *ergodic*. Every column-stochastic matrix has a particularly important eigenvalue:

Lemma 9.14 (Stochastic Spectral Radius). *For any stochastic matrix P:*

1. *1 is an eigenvalue of P.*
2. *The spectral radius $\rho_P = 1$.*
3. *All other eigenvalues satisfy $|\lambda| \leq 1$.*
4. *If P is irreducible, the eigenvalue 1 is simple (algebraic multiplicity one). If P is ergodic (irreducible and aperiodic), then 1 is the unique eigenvalue of P with magnitude 1. If P is irreducible but periodic with period $h > 1$, there are h distinct eigenvalues on the unit circle.*

Proof. First, observe that $\mathbf{1} = (1, 1, \ldots, 1)^T$ is an eigenvector of P^T with eigenvalue 1, since each row of P^T (column of P) sums to 1:

$$P^T \mathbf{1} = \mathbf{1}$$

Since the eigenvalues of P and P^T are identical, this proves that 1 is an eigenvalue of P as well.

For the spectral radius bound, consider any eigenvalue λ of P with eigenvector v. Let i be the index where $|v_i|$ is maximal. Then:

$$\lambda v_i = \sum_{j=1}^{n} p_{ij} v_j$$

Taking absolute values:

$$|\lambda||v_i| = \left| \sum_{j=1}^{n} p_{ij} v_j \right| \leq \sum_{j=1}^{n} p_{ij} |v_j| \leq |v_i| \sum_{j=1}^{n} p_{ij} = |v_i|$$

Since $|v_i| > 0$, we have $|\lambda| \leq 1$, establishing that $\rho_P = 1$. The properties of eigenvalue 1 stated in point 4 are standard results from Perron-Frobenius theory adapted for stochastic matrices. $\square$

A distribution π satisfying $P\pi = \pi$ is called *stationary* – it remains unchanged under the transition rules. For irreducible chains, such a distribution is unique. For ergodic chains, iteration from any initial distribution converges to this stationary state:

Theorem 9.15 (Markov Convergence). *Let P be an ergodic stochastic matrix (i.e., irreducible and aperiodic). Then:*

1. *There exists a unique probability vector π such that $P\pi = \pi$. This π is called the* stationary distribution.

2. *For any initial probability vector $x(0)$:*

$$\lim_{k \to \infty} P^k x(0) = \pi$$

3. *The rate of convergence is governed by the magnitude of the second-largest eigenvalue (i.e., the largest $|\lambda|$ such that $|\lambda| < 1$).*

If P is irreducible but periodic, $P^k x(0)$ does not converge to a single vector but may exhibit periodic limits or converge in Cesaro mean to π.

Example 9.16 (Weather Equilibrium). Returning to our weather model, the matrix P is irreducible (all entries are positive) and aperiodic (e.g., $p_{11} > 0$). Solving $P\pi = \pi$ subject to $\sum \pi_i = 1$ yields stationary distribution $\pi \approx (0.455, 0.295, 0.250)^T$. Thus in the long run, regardless of initial conditions, we expect about 45.5% sunny days, 29.5% cloudy, and 25% rainy. The eigenvalues of P are $1, \approx 0.422, \approx 0.178$. The second-largest eigenvalue magnitude ≈ 0.422 indicates rapid convergence to this equilibrium. $\diamond$

The convergence guaranteed by Theorem 9.15 explains numerous natural phenomena. Population genetics models using stochastic matrices predict allele frequencies in large populations. Economic models of brand loyalty forecast market share evolution. Web page rankings emerge from random walk models of user browsing patterns. In each case, the mathematics of probability conservation shapes long-term behavior through eigenstructure.

Example 9.17 (Population Genetics). Consider a population with two alleles A and a at a single genetic locus. The frequency p of allele A in the parental generation determines offspring genotype probabilities (AA, Aa, aa) under random mating through transition matrix:

$$P = \begin{bmatrix} p^2 & p^2 & p^2 \\ 2p(1-p) & 2p(1-p) & 2p(1-p) \\ (1-p)^2 & (1-p)^2 & (1-p)^2 \end{bmatrix}$$

where Aa and aA represent the same heterozygous genotype. This matrix has identical columns – each offspring's genotype depends only on the overall allele frequencies, not its parents' specific genotypes. The unique stationary distribution gives Hardy-Weinberg equilibrium proportions $(p^2, 2p(1-p), (1-p)^2)$, achieved in just one generation due to the rank-1 structure of P. $\diamond$

Special structures within stochastic matrices reveal additional features. A matrix P is *doubly stochastic* if both its columns and rows sum to one – probability is conserved in both forward and backward time. Such matrices arise naturally in physical systems where transitions conserve some underlying quantity. Their stationary distribution must be uniform: $\pi_i = 1/n$ for all i.

Example 9.18 (Random Walk on a Graph). Consider a particle moving randomly on an undirected graph with n vertices, where at each step it moves with equal probability to any adjacent vertex. The transition probabilities form a stochastic matrix $P = [p_{ij}]$ where

$$p_{ij} = \begin{cases} 1/\deg(j) & \text{if vertices } i \text{ and } j \text{ are adjacent} \\ 0 & \text{otherwise} \end{cases}$$

Here $\deg(j)$ denotes the degree (number of neighbors) of vertex j. This choice makes P column-stochastic, though not necessarily symmetric. However, the relationship $\deg(j)p_{ij} = \deg(i)p_{ji}$ holds due to the undirected nature of the graph.

For instance, consider a graph with six vertices as shown. The transition matrix is:

$$P = \begin{bmatrix} 0 & 1/3 & 1 & 1/2 & 1/3 & 0 \\ 1/4 & 0 & 0 & 1/2 & 1/3 & 0 \\ 1/4 & 0 & 0 & 0 & 0 & 0 \\ 1/4 & 1/3 & 0 & 0 & 0 & 0 \\ 1/4 & 1/3 & 0 & 0 & 0 & 1 \\ 0 & 0 & 0 & 0 & 1/3 & 0 \end{bmatrix}$$

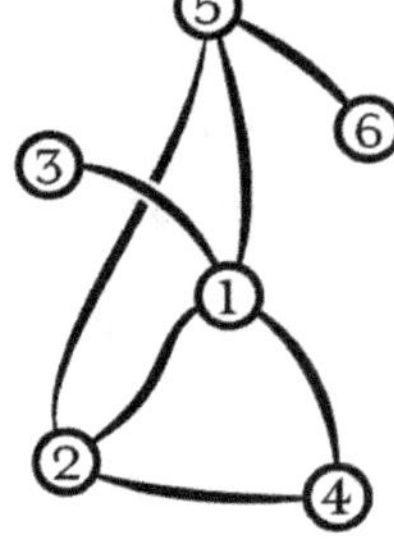

If this graph is connected and not bipartite (which implies aperiodicity for this type of random walk), the chain is ergodic. The stationary distribution has components proportional to vertex degrees: $\pi_1 = 4/14$, $\pi_2 = 3/14$, $\pi_3 = 1/14$, $\pi_4 = 2/14$, $\pi_5 = 3/14$, $\pi_6 = 1/14$. This non-uniform distribution reflects how the random walk spends more time at higher-degree vertices, a principle that underlies many network centrality measures. In particular, observe how vertex 1, with its high degree of 4, captures the largest share of the stationary probability.   ◇

The theory extends naturally to infinite state spaces, though technical care is needed. Random walks on infinite graphs lead to rich theories of recurrence and transience, while scaling limits yield Brownian motion and stochastic differential equations. Throughout, the interplay between probability conservation and matrix structure guides our understanding of how randomness evolves into regularity through linear transformation.

9.5   Symmetric Matrices & Spectra

Mathematics often reveals itself through the relationship between a structure's outer form and inner essence. Like a genome encoding an organism's shape and function, certain morphological features of matrices – outward patterns in their array of numbers – indicate hidden spectral properties that determine this fundamental behavior. The simple structural condition of symmetry ($A^T = A$) yields spectral consequences of profound significance.

Consider, for example, the matrices

$$\begin{bmatrix} 4 & 1 & 2 \\ 1 & 3 & -1 \\ 2 & -1 & 5 \end{bmatrix} \quad \text{or} \quad \begin{bmatrix} 3 & -2 & 0 \\ -2 & 5 & -1 \\ 0 & -1 & 4 \end{bmatrix}$$

Though their entries differ markedly, they share the crucial feature of symmetry about their diagonals. This visible structure – like the bilateral symmetry of a butterfly's wings – reveals deeper patterns: both matrices share spectral properties that emerge not from the specific numbers but from the symmetric pattern itself.

Theorem 9.19 (Spectral Theorem). *Let A be a real symmetric matrix. Then:*

1. *All eigenvalues of A are real*
2. *Eigenvectors corresponding to distinct eigenvalues are orthogonal*
3. *A has an orthonormal basis of eigenvectors*

Thus A can be orthogonally diagonalized: $A = Q\Lambda Q^T$ where Q is orthogonal and Λ is diagonal with real entries.

This remarkable result – that symmetry forces reality of eigenvalues and orthogonality of eigenvectors – transforms abstract matrices into concrete geometric objects. Each symmetric matrix acts by stretching or compressing space along perpendicular axes determined by its eigenvectors. The eigenvalues measure the scale of this deformation, providing a coordinate-independent description of the transformation's action.

The quadratic form associated with symmetric A reveals another face of this structure:

$$q(x) = x^T A x$$

This scalar function measures a kind of generalized "energy" in direction x. Though more abstract than physical energy, this measure guides our understanding of how symmetric matrices shape the spaces they act upon. The following examples illuminate this connection between structure and spectrum across diverse applications.

Example 9.20 (Correlation Structure). Given n measurements of d variables, their correlation matrix $[R] = [\text{CORR}_{ij}]$ records standardized relationships between pairs of variables:

$$\text{CORR}_{ij} = \frac{\sum_{k=1}^{n}(x_{ki} - \overline{x}_i)(x_{kj} - \overline{x}_j)}{\sqrt{\sum_{k=1}^{n}(x_{ki} - \overline{x}_i)^2 \sum_{k=1}^{n}(x_{kj} - \overline{x}_j)^2}}$$

This matrix is naturally symmetric ($\text{CORR}_{ij} = \text{CORR}_{ji}$) with ones on the diagonal ($\text{CORR}_{ii} = 1$). Its spectral structure reveals fundamental patterns in the data:

- Eigenvalues measure strength of correlation patterns
- Eigenvectors identify groups of correlated variables
- Small eigenvalues indicate redundancy or dependencies

For instance, in financial data, eigenvectors often separate market sectors while eigenvalues measure sector-wide versus company-specific variations. In gene expression data, eigenvectors might identify co-regulated genes while eigenvalues quantify the strength of regulatory patterns.   ◇

Example 9.21 (Solid Body Mechanics). The mass of a complex three-dimensional object can be characterized through its inertia matrix (a mass-weighted) covariance matrix:

In mechanics, this is often called the *inertia tensor*. The intuitive fact that asymmetric massive bodies have three "natural" axes of rotation that are all orthogonal is a result of the spectral theorem.

$$\mathcal{I} = \begin{bmatrix} I_{xx} & I_{xy} & I_{xz} \\ I_{xy} & I_{yy} & I_{yz} \\ I_{xz} & I_{yz} & I_{zz} \end{bmatrix}$$

where entries measure second moments of the object's mass or point distribution. The eigenstructure of $\mathcal{I}$ provides a coordinate-independent description:

- Eigenvectors give principal axes of the shape
- Eigenvalues measure spatial extent along these axes
- Ratios of eigenvalues quantify deviation from spherical symmetry

This spectral fingerprint enables shape matching and classification without requiring explicit alignment of objects. A drinking glass has one large eigenvalue (along its length) and two smaller equal ones (across its circular cross-section); a book has three distinct eigenvalues reflecting its rectangular proportions.   ◇

Think: Given only pairwise distances between points, can we reconstruct their relative positions? This inverse problem is akin to trying to deduce the shape of a molecule from measurements between its atoms, or inferring a social network's structure from similarities between individuals.

Example 9.22 (Distance Matrices). Consider a collection of n abstract points with only their pairwise distances known. The squared distances form a symmetric matrix $D = [d_{ij}]$ where d_{ij} represents the squared distance between points i and j. Though we cannot visualize these points directly, their geometric structure hides within D.

The centering matrix $H = I - \frac{1}{n}[1]$ (where $[1]$ denotes the matrix of all ones) and derived matrix $B = -\frac{1}{2}HDH$ play crucial roles. While D itself may not be positive definite, B is symmetric positive semidefinite and encodes the point cloud's geometric essence:

- Positive eigenvalues reveal embedding dimensions
- Eigenvectors reconstruct relative positions
- Spectral decay indicates intrinsic dimensionality

This process of extracting coordinates from distances, known as *classical multidimensional scaling*, demonstrates how eigenstructure can reveal hidden geometric relationships in abstract data. ◇

The extremal properties of quadratic forms associated with symmetric matrices provide another perspective on their structure:

Lemma 9.23 (Extreme Values). *For symmetric A, the quadratic form $q(x) = x^T A x$ achieves:*

1. *Maximum value $\lambda_{\max}$ in direction of largest eigenvalue*
2. *Minimum value $\lambda_{\min}$ in direction of smallest eigenvalue*
3. *Intermediate values between $\lambda_{\min}$ and $\lambda_{\max}$ elsewhere*

on the unit sphere $\|x\| = 1$.

This optimization principle explains why many iterative processes converge toward eigenvectors. The power method for computing dominant eigenvalues, for instance, follows naturally: repeated multiplication by A amplifies components along the direction of largest eigenvalue.

More sophisticated algorithms like Lanczos iteration exploit the connection between symmetry and orthogonality to compute multiple eigenvalues efficiently.

Example 9.24 (Portfolio Optimization). Consider selecting investment weights $w = (w_1, \ldots, w_n)^T$ for n assets, where w_i represents the fraction of wealth invested in asset i. These weights must satisfy two fundamental constraints: they must sum to one (full investment of available funds) and typically must be non-negative (no short selling):

$$\sum_{i=1}^{n} w_i = 1 \quad \text{and} \quad w_i \geq 0$$

The risk of this investment portfolio depends on how the assets' returns move together, captured by their *covariance matrix* $[C] = [C_{ij}]$. Each entry C_{ij} measures how returns of assets i and j vary together – positive values indicate they tend to rise and fall together, while negative values suggest opposite movements.

This covariance matrix is naturally symmetric ($C_{ij} = C_{ji}$), connecting our financial problem to the theory of symmetric matrices. The total portfolio risk, measured by its variance, takes the form of a quadratic expression $w^T[C]w$. In the simplest formulation, our optimization problem becomes:

$$\text{minimize} \quad w^T[C]w$$
$$\text{subject to} \quad w^T\mathbf{1} = 1$$

where $\mathbf{1}$ denotes the vector of all ones. Despite its simple appearance, this constrained minimization reveals surprising depth when analyzed through the Lagrangian:

$$\mathcal{L}(w, \lambda) = w^T[C]w - \lambda(w^T\mathbf{1} - 1)$$

Setting $\nabla_w \mathcal{L} = \mathbf{0}$ yields:

$$2[C]w - \lambda\mathbf{1} = \mathbf{0}$$

Recall: For minimizing $F(x)$ subject to $G(x) = 0$, the Lagrangian $\mathcal{L}(x, \lambda) = F(x) - \lambda G(x)$ yields equations $\nabla_x \mathcal{L} = \mathbf{0}$ and $G(x) = 0$. Here $F(w) = w^T[C]w$ and $G(w) = w^T\mathbf{1} - 1$.

Together with the constraint equation, this system determines the optimal weights:

$$\begin{aligned} w &= \tfrac{1}{2}[C]^{-1}\mathbf{1}\lambda \\ 1 &= w^T\mathbf{1} = \tfrac{1}{2}\lambda\mathbf{1}^T[C]^{-1}\mathbf{1} \end{aligned}$$

The solution reveals a fundamental relationship:

$$w = \frac{[C]^{-1}\mathbf{1}}{\mathbf{1}^T[C]^{-1}\mathbf{1}}$$

Think: The constraint $w^T\mathbf{1} = 1$ defines a hyperplane in $\mathbb{R}^n$. The solution occurs where level sets of the quadratic form become tangent to this hyperplane – a geometric picture explaining why constrained optimization often yields such elegant solutions.

This transformation – from minimizing risk under investment constraints to solving a linear system – exemplifies a deeper pattern. Many constrained optimization problems involving symmetric matrices reduce to such systems through the machinery of Lagrange multipliers. The eigenstructure of $[C]$ shapes both the optimal portfolio weights and the minimum achievable risk. In practice, this analysis extends naturally to include expected returns through the Markowitz mean-variance framework, where risk minimization balances against return maximization through careful exploitation of covariance structure.   ◇

This intimate relationship between structural symmetry and spectral properties exemplifies a deeper pattern in mathematics: morphological constraints often encode spectral essence. Just as an organism's form reflects its genome, a matrix's visible patterns encode hidden properties that determine its fundamental behavior. This theme will deepen as we explore structured matrices in network science and develop the singular value decomposition in Chapter 10.

9.6 Networked Behavior & Consensus

The flow of influence through networks shapes everything from opinion formation to economic behavior to artificial intelligence. When agents in a network update their states based on their neighbors' values, complex global patterns can emerge from simple local rules. Understanding such collective behavior requires uniting the spectral theory of Sections 9.3–9.5 with the concrete topology of interaction networks.

Definition 9.25 (Graph Laplacian). For an undirected graph with n vertices, the *graph Laplacian* $L = [L_{ij}]$ is an $n \times n$ matrix whose entries are:

$$L_{ij} = \begin{cases} d_i & \text{if } i = j \\ -1 & \text{if vertices } i \text{ and } j \text{ are connected} \\ 0 & \text{otherwise} \end{cases}$$

where d_i denotes the *degree* of vertex i – the number of edges connected to it. Equivalently, if $D = \text{DIAG}(d_1, \ldots, d_n)$ is the *degree matrix* and $A = [a_{ij}]$ is the *adjacency matrix* with $a_{ij} = 1$ for connected vertices and 0 otherwise, then $L = D - A$. •

This matrix, though simple to define, carries profound information about network topology through its spectrum. Its action on a vector measures how quantities diffuse across edges, making it fundamental to understanding collective dynamics.

Consider a network of n agents, each holding a real value $x_i(t)$ that evolves in discrete time through local averaging:

$$x_i(t+1) = \sum_{j \in N(i)} w_{ij} x_j(t)$$

where $N(i)$ denotes the neighbors of agent i and weights w_{ij} represent interaction strengths. Writing $x(t)$ for the vector of states, this local update rule becomes matrix iteration:

$$x(t+1) = Wx(t)$$

The weight matrix $W = [w_{ij}]$ often takes the form $W = D^{-1}A$ (neighbors weighted equally) or $W = I - \epsilon L$ (gradient descent on disagreement), connecting back to our study of stochastic matrices in Section 9.4.

Example 9.26 (Opinion Dynamics). Consider a political discourse network where there are five rough cohorts of people on social media, with connections as shown (these are not individuals, but rather interacting groups with aggregate opinions). Groups 1 and 4 hold the most influential positions, each connected to three others. Each group updates their opinion on a political proposal based on a weighted average of their neighbors' views:

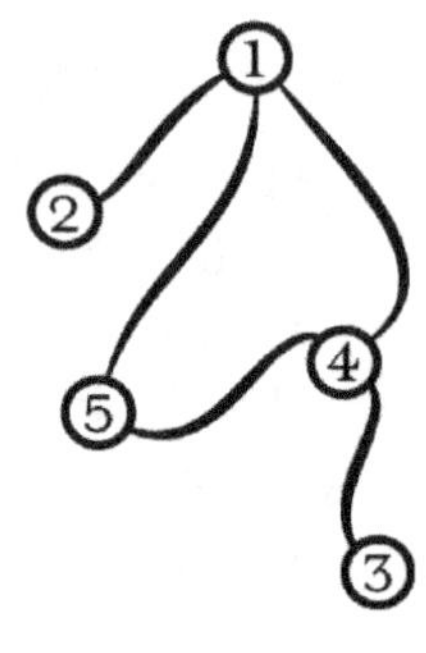

$$W = \begin{bmatrix} 1/4 & 1/2 & 0 & 1/4 & 1/3 \\ 1/4 & 1/2 & 0 & 0 & 0 \\ 0 & 0 & 1/2 & 1/4 & 0 \\ 1/4 & 0 & 0 & 1/4 & 1/3 \\ 1/4 & 0 & 1/2 & 1/4 & 1/3 \end{bmatrix}$$

The column-stochasticity of W ensures each new opinion distribution remains a proper probability distribution. Starting from polarized initial opinions $x(0) = (1, -1, -1, 1, -1)^T$, iteration reveals steady convergence toward consensus, with the well-connected groups exercising greater influence on the final value through their multiple pathways of influence.

◇

When the network is connected (the matrix W is irreducible as in Section 9.3) and the weights respect certain balance conditions, a remarkable phenomenon occurs: all agents converge to consensus. This convergence follows directly from the Perron-Frobenius theory of Section 9.3, but the network perspective reveals how topology shapes the path to agreement.

Theorem 9.27 (Network Consensus). *Let W be an $n \times n$ column-stochastic matrix that is also ergodic (i.e., irreducible and aperiodic). Let π be the unique stationary probability distribution for W (the eigenvector corresponding to eigenvalue $\lambda_1 = 1$, normalized such that $\sum_i \pi_i = 1$). Then for any initial state vector $x(0) \in \mathbb{R}^n$:*

$$\lim_{t \to \infty} W^t x(0) = \left(\sum_{i=1}^{n} x_i(0) \right) \pi$$

Moreover, the rate of convergence to this consensus state (or its scaled version) is determined by $|\lambda_2|$, where λ_2 is an eigenvalue of W with the second-largest magnitude (so $|\lambda_2| < 1$). The convergence is geometric, with the error typically decreasing by a factor of approximately $|\lambda_2|$ at each step.

The term $\sum x_i(0)$ is the sum of the initial states. If $x(0)$ is itself a probability distribution, this sum is 1, and the system converges to π. If W is only irreducible and periodic, $W^t x(0)$ does not generally converge to a single vector.

This theorem connects network structure to collective behavior through spectral properties. The entries of π represent the relative influence of each node in the final consensus state, while $|\lambda_2|$ measures how quickly agreement emerges. Highly connected networks with good mixing properties have large spectral gaps and rapid convergence, as we saw with stochastic matrices in Section 9.4.

Example 9.28 (Asset Price Formation). Consider a market network where traders adjust their valuation of an asset based on their trading partners' prices. The weight matrix W reflects both trust relationships and trading volume between parties. The consensus price that emerges represents a form of market efficiency, with well-connected traders (high π_i values) having greater influence on price discovery. The spectral gap $1 - |\lambda_2|$ determines how quickly the market reaches this equilibrium – critical for understanding market stability and reaction to shocks. ◇

Example 9.29 (Robotic Flocking). A swarm of robots moving in the plane can achieve coordinated motion through local averaging of velocities. Each robot i has position p_i and velocity v_i, updating according to:

$$v_i(t+1) = v_i(t) + \epsilon \sum_{j \in N_i(t)} (v_j(t) - v_i(t))$$

where $N_i(t)$ contains robots within communication range of robot i. This can be written as iteration with a time-varying Laplacian:

$$v(t+1) = (I - \epsilon L(t))v(t)$$

The spectrum of $L(t)$ determines both stability of the flocking behavior and convergence rate to aligned motion. This example shows how the tools developed for static networks extend naturally to dynamic topologies. ◇

Beyond simple averaging, the graph Laplacian enables more sophisticated analysis through its connection to network structure. For a connected graph:

1. The zero eigenvalue is simple, with eigenvector $\mathbf{1}$
2. The smallest nonzero eigenvalue measures connectivity
3. The corresponding eigenvector reveals natural network clusters

These properties, emerging from the symmetry studied in Section 9.5, provide both analytical tools and design principles for engineering collective behavior.

Example 9.30 (Supply Chain Networks). Returning to the input-output model from Section 9.1, the Laplacian spectrum reveals vulnerability to disruption. Industries corresponding to components of the Fiedler vector (the eigenvector of the smallest nonzero eigenvalue) tend to split first under stress, identifying natural fault lines in the economic network. The magnitude of this eigenvalue quantifies the network's robustness to such splits. ◇

Modern applications of these principles abound in both technology and nature:

- Social networks shape opinion formation and information spread
- Financial networks transmit shocks and determine systemic risk
- Robot swarms coordinate through local interactions
- Supply chains balance efficiency against resilience

In each case, the interplay between network structure and spectral properties determines system-level behavior.

The tools we have developed – from stochastic matrices to symmetric spectra – unite in the study of network dynamics. This synthesis reveals how local rules and global topology combine to produce collective intelligence, pointing toward both deeper mathematics and practical applications in the chapters ahead. The final application of this chapter will explore one particularly elegant application: the PageRank algorithm that grew from these principles to organize the early Web.

Spectral Graph Theory & Vibrations

The mathematics of musical instruments emerges through the vibrations of strings and membranes. A plucked string settles into standing waves whose frequencies determine musical pitch, while a drum head vibrates in complex patterns creating its characteristic timbre. Though seemingly far from the matrix theory developed in this chapter, these physical systems find their mathematical essence in the eigenvalues of *graph Laplacians* — a connection that transforms discrete network analysis into continuous harmonics through the deep principles of spectral theory.

Consider first a string fixed at its endpoints, discretized into n points connected by identical springs. The displacement x_i of point i from equilibrium feels restoring forces from its neighbors proportional to the displacement differences:

$$m\frac{d^2 x_i}{dt^2} = \kappa(x_{i+1} - x_i) - \kappa(x_i - x_{i-1})$$

where m is the mass of each point and κ the spring constant. Written in matrix form:

$$m\frac{d^2 x}{dt^2} = -\kappa L x$$

where L is the *graph Laplacian*:

$$L = \begin{bmatrix} 1 & -1 & 0 & \cdots & 0 \\ -1 & 2 & -1 & \cdots & 0 \\ 0 & -1 & 2 & \cdots & 0 \\ \vdots & \vdots & \vdots & \ddots & \vdots \\ 0 & 0 & 0 & \cdots & 1 \end{bmatrix}$$

Note: The graph Laplacian is symmetric positive semidefinite, with eigenvalues directly determining vibrational frequencies.

The eigenvectors of L represent standing wave patterns — modes of vibration where all points move sinusoidally with the same frequency but different amplitudes. The corresponding eigenvalues λ_i determine these frequencies through $\omega_i = \sqrt{k\lambda_i/m}$. As n increases, these discrete modes converge to the continuous solutions:

$$v_k(x) = \sin\left(\frac{k\pi x}{L}\right), \quad \omega_k = \frac{k\pi}{L}\sqrt{\frac{T}{\rho}}$$

where L is string length, T tension, and ρ linear density.

Think: The convergence of discrete to continuous eigenmodes reveals how graph theory naturally extends to continuous systems.

This connection between graph spectra and physical vibrations extends to higher dimensions. For a drum head approximated by a triangular mesh, focus on the vertices and edges of the mesh and define the graph Laplacian as a square matrix on the vertex set V:

$$L_{ij} = \begin{cases} \deg(i) & \text{if } i = j \\ -1 & \text{if } i \sim j \\ 0 & \text{otherwise} \end{cases}$$

where $\deg(i)$ counts the number of vertices adjacent to vertex i, and $i \sim j$ denotes adjacent vertices. The eigenvalues again determine vibrational frequencies, while eigenvectors describe mode shapes — patterns of displacement that maintain their form while oscillating in time.

Example: A circular drum head's modes form the remarkable *Bessel functions*, emerging as limits of mesh eigenvectors.

A remarkable result connects these spectra to pure graph theory: the multiplicity of the Laplacian's zero eigenvalue equals the number of connected components. This reveals an elegant duality — just as low eigenvalues describe slow vibrations, they also capture fundamental topological properties. The second-smallest eigenvalue measures how well-connected the graph is, with larger values indicating stronger connectivity, much as dominant eigenvalues determined convergence rates in our study of stochastic matrices.

Nota bene: The connection between eigenvalues and graph connectivity parallels how dominant eigenvalues controlled network convergence earlier in this chapter.

The spectral approach to graph analysis extends naturally to partitioning problems, where we seek to divide vertices into groups with many internal connections but few connections between groups. The eigenvector corresponding to the second-smallest eigenvalue provides a continuous relaxation of this discrete problem: partition vertices according to the sign of their corresponding entry. This spectral clustering method often reveals natural communities in networks.

For weighted graphs, where edges carry different strengths, define the degree matrix $D = \mathrm{DIAG}(d_1, \ldots, d_n)$ where $d_i = \sum_j w_{ij}$ sums the weights of edges incident to vertex i. The normalized Laplacian $\mathcal{L} = D^{-1/2} L D^{-1/2}$ accounts for these varying connection strengths while maintaining symmetry. Its eigenvalues lie in $[0, 2]$, providing a normalized measure of graph structure independent of absolute edge weights.

The deep connection between discrete graphs and continuous vibrations reveals a profound unity in mathematical physics. Eigenvalues and eigenvectors encode both physical oscillations and abstract graph properties, while the Laplacian provides the bridge between structure and dynamics. This synthesis — of the discrete and continuous through spectral theory — exemplifies how fundamental mathematical principles illuminate seemingly disparate phenomena.

PageRank: The Flow of Web Authority

The World Wide Web presents perhaps the largest human-constructed network in history — billions of pages connected through hyperlinks that channel attention and information across the digital sphere. Like the discrete-time systems studied in Section 9.1, this vast network exhibits intrinsic patterns of information flow that can be understood through careful mathematical analysis. The challenge of organizing this space led to one of the most elegant applications of stochastic matrices and iterative convergence: Google's PageRank algorithm.

Consider a web surfer following links from page to page, modeling their behavior as the type of Markov chain developed in Section 9.4. At each step, they either follow a randomly chosen outgoing link (with probability α) or jump to a random page anywhere on the web (with probability $1 - \alpha$). This process generates a *transition matrix* $P = [p_{ij}]$:

$$p_{ij} = \alpha \frac{a_{ij}}{d_j} + \frac{1 - \alpha}{n}$$

where $a_{ij} = 1$ if page j links to page i (and 0 otherwise), d_j is the number of outgoing links from page j (if $d_j = 0$, this term is often handled by assuming jumps to all pages equally), and n is the total number of pages. The term $(1 - \alpha)/n$ represents the random teleportation probability.

This construction makes P a column-stochastic matrix. Since $0 < \alpha < 1$ and $n > 0$, all entries p_{ij} are strictly positive. A strictly positive stochastic matrix is ergodic. Applying the theory developed in Section 9.4 (specifically Theorem 9.15), we know this matrix P must have a unique stationary probability distribution π satisfying $\pi = P\pi$. This *PageRank vector* π measures each page's importance through its long-term visit probability.

Just as with the consensus problems studied in Section 9.6, the computation of π naturally employs iterative methods (the power method). Starting from any initial probability distribution x_0 (typically uniform), repeated multiplication by P converges to π:

$$x_{k+1} = Px_k \quad \text{and} \quad \lim_{k \to \infty} x_k = \pi$$

The properties of P ensure each iterate remains a probability distribution and that convergence to the unique limit π occurs.

Example 9.31 (Simple Web). Consider a tiny web of four pages with link structure given by adjacency matrix A. With damping factor $\alpha = 0.85$, the iteration converges to:

$$\pi \approx \begin{pmatrix} 0.17 \\ 0.31 \\ 0.35 \\ 0.17 \end{pmatrix}$$

This distribution reflects both local link structure and global network position. Like the network centrality measures from Section 9.6, pages with more incoming paths tend to receive higher scores. ◇

The rate of convergence, as with all ergodic Markov chains, is determined by the magnitude of the second-largest eigenvalue of P, denoted $|\lambda_2(P)|$. The

Historical Note: The addition of random teleportation (typically $\alpha \approx 0.85 < 1$) ensures that P is a *primitive* matrix if $n > 0$: all its entries are strictly positive. A primitive stochastic matrix is necessarily *ergodic* (irreducible and aperiodic), guaranteeing the convergence properties studied in Section 9.4.

eigenvalues of P are related to those of the normalized adjacency matrix part by $\lambda_k(P) = \alpha\lambda_k(M_{norm}) + (1 - \alpha)/n$ for the eigenvector corresponding to $\lambda_k(M_{norm})$ if it's orthogonal to the all-ones vector, and $\lambda_1(P) = 1$. Thus, $|\lambda_2(P)| \approx \alpha|\lambda_2(M_{norm})|$. For typical web graphs and $\alpha \approx 0.85$, $|\lambda_2(P)|$ is often close to α.

Example 9.32 (Convergence Behavior). For our four-page example, tracking successive iterates reveals geometric convergence:

$$\|x_k - \pi\| \approx |\lambda_2(P)|^k\|x_0 - \pi\|$$

where $|\lambda_2(P)| < 1$. For $\alpha = 0.85$, this value is typically close to 0.85, leading to reasonably fast convergence. This matches the behavior predicted by our Markov chain analysis. ◇

The web graph provides a perfect example of the network structures studied in Section 9.6, where topology shapes dynamic behavior. PageRank's success stems from how it unites two key principles from our chapter: the convergence of ergodic stochastic matrices and the flow of influence through networks. This synthesis — of Markov chain theory with network dynamics — transformed web search through pure mathematics.

The framework extends naturally to other contexts where importance flows along network edges, provided the construction ensures an ergodic transition matrix for unique convergence:

- Citation networks ranking academic papers
- Social networks measuring user influence
- Economic networks revealing systematic importance

PageRank exemplifies how the mathematics developed in this chapter shapes the modern world. The stochastic matrices, network dynamics, and iterative methods we have studied combine to organize the chaotic web through rigorous theory. This transformation of abstract mathematics into practical computation demonstrates the profound utility of careful mathematical analysis in contemporary engineering.

Exercises: Chapter 9

1. Let $A = \begin{bmatrix} 0 & 1 & 0 \\ 1 & 0 & 1 \\ 0 & 1 & 0 \end{bmatrix}$. Compute the first five powers of A and describe any patterns you observe. What does this tell you about random walks on the corresponding graph?

2. For the transition matrix $P = \begin{bmatrix} 0.5 & 0.3 & 0.2 \\ 0.4 & 0.4 & 0.3 \\ 0.1 & 0.3 & 0.5 \end{bmatrix}$, compute P^2 and P^3. What appears to be happening to the powers of P? Explain this behavior using the theory of stochastic matrices.

3. Consider a stochastic matrix $P = \begin{bmatrix} 0.8 & 0.2 \\ 0.3 & 0.7 \end{bmatrix}$. Find its stationary distribution. Is

this distribution unique? Explain why or why not using the Perron-Frobenius theory.

4. Let G be an undirected graph with vertices $\{1,2,3,4\}$ and edges

$$E = \{(1,2),(2,3),(3,4),(4,1),(2,4)\}$$

Write down its adjacency matrix A and degree matrix D; then compute the graph Laplacian $L = D - A$ and find its eigenvalues. What do these tell you about the graph's connectivity?

5. A gambler starts with \$2 and plays a game where they win \$1 with probability $1/3$ and lose \$1 with probability $2/3$. They stop playing when they either go broke (\$0) or reach their goal of \$3. Model this as a Markov chain with states $\{0,1,2,3\}$. Find the transition matrix and compute the probability of eventually reaching the goal state.

6. A child's toy has three buttons: red, blue, and green. When pressed, each button plays a tune and randomly transitions to lighting up another button (including possibly the same one) according to transition matrix

$$P = \begin{bmatrix} 0.2 & 0.4 & 0.3 \\ 0.5 & 0.3 & 0.4 \\ 0.3 & 0.3 & 0.3 \end{bmatrix}$$

If the red button is pressed first, what is the probability that the green button will be lit after exactly three presses? After many presses, what fraction of time will each button be lit?

7. Consider an input-output economic model with three sectors where the input matrix is $A = \begin{bmatrix} 0.3 & 0.2 & 0.1 \\ 0.2 & 0.4 & 0.3 \\ 0.1 & 0.2 & 0.4 \end{bmatrix}$. Find the equilibrium production levels. What does the Perron-Frobenius theorem tell you about the stability of this economy?

8. Let A be the adjacency matrix of a simple undirected graph. Prove that the (i,j) entry of A^k counts the number of walks of length k from vertex i to vertex j.

9. A matrix P is doubly stochastic if both its rows and columns sum to 1. Prove that if P is doubly stochastic, then the vector $\mathbf{1}/n$ (where n is the dimension) must be a stationary distribution. Is this distribution necessarily unique?

10. For a connected undirected graph, the random walk Laplacian is defined as $L_{rw} = I - D^{-1}A$ where D is the degree matrix and A is the adjacency matrix. Prove that L_{rw} always has eigenvalue 0 with eigenvector $\mathbf{1}$.

11. Let P be an irreducible stochastic matrix. Prove that if P has an eigenvalue λ with $|\lambda| = 1$, then λ must be a root of unity (i.e., $\lambda^k = 1$ for some positive integer k).

12. Consider a network of 4 agents where each agent updates their opinion based on a weighted average of their neighbors' opinions. If the network is a square (4-cycle), write down the update matrix and determine how quickly opinions will converge to consensus.

13. Let G be a graph and L its Laplacian matrix. Prove that the multiplicity of the eigenvalue 0 equals the number of connected components in G.

14. For an irreducible stochastic matrix P, show that $\|P^n x\| \leq \|x\|$ for any vector x perpendicular to $\mathbf{1}$, where $\|\cdot\|$ is the standard Euclidean norm.

15. Let P be a stochastic matrix representing transition probabilities in a Markov chain. Define the *mean first passage time* m_{ij} as the expected number of steps to reach state j starting from state i. Show that these times satisfy the equation $m_{ij} = 1 + \sum_{k \neq j} p_{ki} m_{kj}$ for $i \neq j$.

16. Consider a Markov chain with transition matrix P and suppose state j is *absorbing* (meaning $p_{jj} = 1$ and $p_{ij} = 0$ for $i \neq j$). Prove that if the chain starts in state $i \neq j$, the probability of eventual absorption in state j equals the (i, j) entry of $(I - Q)^{-1}R$, where Q is P with row and column j removed, and R is column j of P with entry j removed.

17. Let P be the transition matrix of an irreducible Markov chain, and let π be its stationary distribution. The *time-reversed* chain has transition matrix $\tilde{P}$ where

$$\tilde{p}_{ij} = \frac{\pi_i}{\pi_j} p_{ji}$$

Prove that $\tilde{P}$ is stochastic and shares the same stationary distribution as P. What does this tell you about the reversibility of the chain?

18. The entropy of a probability distribution π is defined as $H(\pi) = -\sum_i \pi_i \ln(\pi_i)$. For a stochastic matrix P, prove that if π is a stationary distribution, then π maximizes the entropy among all distributions x satisfying $Px = x$.

19. Consider the graph of a cube (8 vertices, 12 edges). Write down its adjacency matrix and compute its spectrum. What does the spectrum tell you about random walks on this graph?

20. The *clustering coefficient* of a vertex i in an undirected graph is defined as the fraction of pairs of i's neighbors that are themselves connected. Given the adjacency matrix of a graph, write out the formula for computing the clustering coefficient of a vertex in terms of matrix operations. Apply your formula to compute the clustering coefficient of each vertex in the graph with adjacency matrix:

$$A = \begin{bmatrix} 0 & 1 & 1 & 0 \\ 1 & 0 & 1 & 1 \\ 1 & 1 & 0 & 1 \\ 0 & 1 & 1 & 0 \end{bmatrix}$$

21. The *eigenvector centrality* x of nodes in a network is defined as the solution to $\lambda x = Ax$ where A is the adjacency matrix and λ is the spectral radius of A. Given a network represented by adjacency matrix:

$$A = \begin{bmatrix} 0 & 1 & 1 & 0 & 0 \\ 1 & 0 & 1 & 1 & 0 \\ 1 & 1 & 0 & 1 & 1 \\ 0 & 1 & 1 & 0 & 1 \\ 0 & 0 & 1 & 1 & 0 \end{bmatrix}$$

Use power iteration to approximate the eigenvector centrality scores (normalize after each iteration). How do these scores relate to each node's degree? Explain why some nodes have higher centrality than their degree would suggest.

Chapter 10
Singular Value Decomposition

"build we the Mundane Shell around the Rock of Albion"

EVERY TRANSFORMATION HARBORS hidden symmetries beneath its surface complexity. Like crystal structures buried in seemingly formless rock, these patterns reveal themselves only through careful excavation and analysis. The eigendecomposition developed in previous chapters illuminates the structure of square matrices through their action under iteration. Yet this tool, powerful as it is, reaches only part way to the deepest patterns underlying linear transformations. A more fundamental decomposition lies waiting to be unearthed.

The limitation of eigentheory to square matrices is no accident – eigenvalues and eigenvectors emerge naturally from iteration, which requires a transformation to map a space to itself. Yet the geometric essence of a linear transformation – how it stretches and rotates space – extends beyond such endomorphisms. Every linear transformation, whether square or rectangular, admits a canonical decomposition that reveals its intrinsic geometric character. This decomposition exposes not just preferred directions, but fundamental relationships between input and output spaces that remain invisible to eigentheory alone.

Our task is to dig beneath the surface structure of linear transformations to discover these deeper patterns. We begin with the geometric intuition of how matrices transform spheres into ellipsoids, revealing natural input and output directions. From these foundations emerges the singular value decomposition, providing both theoretical insight and practical tools for understanding linear transformations in their full generality.

The power of this decomposition lies in its fusion of algebraic and geometric perspectives. What appears as abstract factorization in coordinates manifests geometrically as an optimal sequence of simple transformations. This optimality principle – that the singular value decomposition provides best possible approximations in various precise senses – transforms our theoretical understanding into practical methods for data compression, signal processing, and dimensionality reduction. The tools we develop here will prove fundamental to modern applications from image processing to machine learning.

10.1   Spheres, Ellipsoids, & Singular Values

The geometry of a linear transformation $A \in \mathbb{R}^{m \times n}$ is vividly revealed by its action on the unit sphere in its domain, $\mathbb{R}^n$. The unit sphere, defined by $\|x\|^2 = 1$ or $x^T x = 1$, is deformed by A into an ellipsoid in the codomain, $\mathbb{R}^m$. Understanding this deformation is key.

The shape and orientation of this output ellipsoid are determined by the symmetric positive semidefinite matrix $A^T A$. The squared length of a transformed vector Ax is given by:

$$\|Ax\|^2 = (Ax)^T(Ax) = x^T(A^T A)x.$$

By the Spectral Theorem (Theorem 9.19), $A^T A$ (being an $n \times n$ symmetric matrix) has n real, non-negative eigenvalues $\lambda_1 \geq \lambda_2 \geq \cdots \geq \lambda_n \geq 0$. Let $V = [v_1 \ldots v_n]$ be an orthogonal matrix whose columns are the corresponding orthonormal eigenvectors of $A^T A$. These eigenvectors v_i represent principal directions in the input space $\mathbb{R}^n$. When x is one of these eigenvectors, say $x = v_i$, then

$$\|Av_i\|^2 = v_i^T(A^T A)v_i = v_i^T(\lambda_i v_i) = \lambda_i \|v_i\|^2 = \lambda_i.$$

Thus, the matrix A stretches the principal input direction v_i by a factor of $\sqrt{\lambda_i}$. These stretching factors are fundamental to the transformation A.

Definition 10.1 (Singular Values). Let $A \in \mathbb{R}^{m \times n}$. The $n \times n$ matrix $A^T A$ is symmetric and positive semidefinite, so its eigenvalues λ_i are real and non-negative. The *singular values* of A, denoted σ_i, are the square roots of these eigenvalues: $\sigma_i = \sqrt{\lambda_i}$. They are conventionally arranged in descending order:

$$\sigma_1 \geq \sigma_2 \geq \cdots \geq \sigma_n \geq 0.$$

The number of non-zero singular values is $r = \operatorname{rank}(A^T A) = \operatorname{rank}(A)$. The corresponding orthonormal eigenvectors v_i of $A^T A$ are called the *right singular vectors* of A.

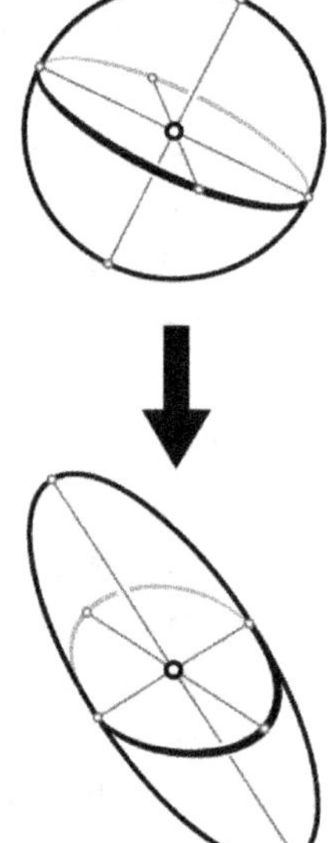

The eigenvalues λ_i specifically refer to those of $A^T A$.

The image of the unit sphere under A is an ellipsoid (possibly degenerate if A is rank-deficient) in $\mathbb{R}^m$. The semi-axes of this ellipsoid are aligned with certain vectors $u_i \in \mathbb{R}^m$ and have lengths equal to the non-zero singular values σ_i. The directions v_i in $\mathbb{R}^n$ are mapped by A to these semi-axis vectors: $Av_i = \sigma_i u_i$. The vectors u_i will be the *left singular vectors*.

Example 10.2 (Image Transformation). Consider the 2×2 matrix:

$$A = \begin{bmatrix} 3 & 1 \\ 1 & 2 \end{bmatrix} \quad \Rightarrow \quad A^T A = \begin{bmatrix} 10 & 5 \\ 5 & 5 \end{bmatrix}$$

The characteristic polynomial of $A^T A$ is $\lambda^2 - 15\lambda + 25 = 0$. Its eigenvalues are $\lambda_1 = (15 + \sqrt{125})/2 \approx 13.09$ and $\lambda_2 = (15 - \sqrt{125})/2 \approx 1.91$. The singular values are $\sigma_1 = \sqrt{\lambda_1} \approx 3.618$ and $\sigma_2 = \sqrt{\lambda_2} \approx 1.382$. The unit circle in $\mathbb{R}^2$ is transformed by A into an ellipse whose semi-axes have lengths σ_1 and σ_2. The directions of these semi-axes in $\mathbb{R}^2$ (the domain) are given by the eigenvectors of $A^T A$. $\diamond$

This geometric picture – mapping principal input directions (eigenvectors of $A^T A$) to scaled principal output directions – forms the intuitive basis for the Singular Value Decomposition.

10.2   Constructing the SVD

The geometric insight that a linear transformation A maps orthonormal principal input directions to orthogonal principal output directions, scaled by singular values, leads directly to its most fundamental factorization. We aim to find orthogonal matrices U and V and a rectangular diagonal matrix Σ such that $A = U\Sigma V^T$.

Step 1: Finding V and the Singular Values σ_i.

As established in Section 10.1, the matrix $A^T A$ is an $n \times n$ symmetric positive semidefinite matrix. By the Spectral Theorem, there exists an $n \times n$ orthogonal matrix $V = [v_1 \dots v_n]$ whose columns are orthonormal eigenvectors of $A^T A$, and an $n \times n$ diagonal matrix $\Lambda = \text{DIAG}(\lambda_1, \dots, \lambda_n)$ of corresponding non-negative eigenvalues, such that $A^T A = V \Lambda V^T$. The singular values of A are $\sigma_i = \sqrt{\lambda_i}$, ordered $\sigma_1 \geq \sigma_2 \geq \cdots \geq \sigma_n \geq 0$. Let r be the rank of A, which is also the number of non-zero singular values. The columns $v_1, \dots, v_r$ form an orthonormal basis for $(\ker A)^\perp = \text{im}(A^T)$, and $v_{r+1}, \dots, v_n$ form an orthonormal basis for $\ker A = \ker(A^T A)$.

Step 2: Defining the Left Singular Vectors U.

For each $i = 1, \ldots, r$ (where $\sigma_i > 0$), define the vector $u_i \in \mathbb{R}^m$ by

$$u_i = \frac{1}{\sigma_i} A v_i.$$

These r vectors are orthonormal. To see this, consider their inner product:

$$u_i^T u_j = \left(\frac{1}{\sigma_i} A v_i\right)^T \left(\frac{1}{\sigma_j} A v_j\right) = \frac{1}{\sigma_i \sigma_j} v_i^T (A^T A) v_j$$

$$= \frac{1}{\sigma_i \sigma_j} v_i^T (\lambda_j v_j) \quad \text{(since } v_j \text{ is an eigenvector of } A^T A\text{)}$$

$$= \frac{\sigma_j^2}{\sigma_i \sigma_j} (v_i^T v_j).$$

Since $\{v_k\}$ is an orthonormal set, $v_i^T v_j = \delta_{ij}$. Thus, $u_i^T u_j = \frac{\sigma_j}{\sigma_i} \delta_{ij}$. For $i = j$, $u_i^T u_i = 1$. For $i \neq j$, $u_i^T u_j = 0$. So, $\{u_1, \ldots, u_r\}$ is an orthonormal set of r vectors in $\mathbb{R}^m$. These vectors form an orthonormal basis for the column space of A, $\text{im}(A)$.

If $r < m$, the set $\{u_1, \ldots, u_r\}$ does not span all of $\mathbb{R}^m$. We can extend it to a full orthonormal basis for $\mathbb{R}^m$ by choosing an additional $m - r$ orthonormal vectors $\{u_{r+1}, \ldots, u_m\}$ that form a basis for $(\text{im } A)^\perp = \ker(A^T)$. Let $U = [u_1 \ldots u_r \ldots u_m]$ be the $m \times m$ orthogonal matrix whose columns are these left singular vectors.

Step 3: Defining the Matrix Σ.

Let Σ be the $m \times n$ matrix whose entry $\Sigma_{ii} = \sigma_i$ for $i = 1, \ldots, p = \min(m, n)$, and all other entries are zero. If $r < p$, then $\sigma_{r+1}, \ldots, \sigma_p$ are zero.

With U, Σ, and V thus constructed, we have $A v_i = \sigma_i u_i$ for $i - 1, \ldots, r$, and $A v_i = 0$ for $i = r + 1, \ldots, n$ (since these v_i are in $\ker A$). This set of vector equations can be written in matrix form as $AV = U\Sigma'$, where Σ' is an $m \times n$ matrix whose first r diagonal entries are $\sigma_1, \ldots, \sigma_r$ and all other entries are zero. This Σ' is precisely our Σ. Since V is orthogonal, $V^{-1} = V^T$, so $AV = U\Sigma$ implies $A = U\Sigma V^T$.

This construction leads to the central theorem:

Theorem 10.3 (Singular Value Decomposition). *Every matrix $A \in \mathbb{R}^{m \times n}$ admits a decomposition*

$$A = U\Sigma V^T \tag{10.1}$$

where:

Nota bene: The columns of U are also the orthonormal eigenvectors of AA^T, and the non-zero eigenvalues of AA^T are $\sigma_1^2, \ldots, \sigma_r^2$, the same as for $A^T A$. One could alternatively start by diagonalizing AA^T to find U and the σ_i^2.

1. $U \in \mathbb{R}^{m \times m}$ is an orthogonal matrix whose columns are the left singular vectors of A.
2. $V \in \mathbb{R}^{n \times n}$ is an orthogonal matrix whose columns are the right singular vectors of A.
3. $\Sigma \in \mathbb{R}^{m \times n}$ is a rectangular diagonal matrix, where the diagonal entries $\Sigma_{ii} = \sigma_i$ are the singular values of A, ordered $\sigma_1 \geq \sigma_2 \geq \cdots \geq \sigma_p \geq 0$ with $p = \min\{m, n\}$.

The singular values σ_i are uniquely determined.

This SVD reveals the fundamental action of A: it maps the i-th right singular vector v_i to σ_i times the i-th left singular vector u_i:

$$Av_i = \sigma_i u_i \quad \text{for } i = 1, \ldots, \min(m, n).$$

If $\sigma_i = 0$, then $Av_i = \mathbf{0}$. The transformation A essentially acts by:

1. Rotating/reflecting the input space so the basis vectors e_i align with v_i (action of V^T).
2. Scaling these aligned vectors by σ_i along the new axes (action of Σ).
3. Rotating/reflecting the result into the output space so the scaled axes align with u_i (action of U).

The SVD is arguably the most important matrix factorization, providing deep insights into a matrix's structure, geometry, rank, and numerical properties. Its applications, explored in subsequent sections and chapters, are vast.

10.3   Interpreting the SVD

The Singular Value Decomposition $A = U\Sigma V^T$ provides far more than a mere algebraic factorization of a matrix $A \in \mathbb{R}^{m \times n}$; it unveils the intrinsic geometry and fundamental structure of the linear transformation A represents. Having constructed the components U, Σ, and V in Section 10.2, we now consider their meaning and how they connect to core concepts like rank, the four fundamental subspaces, and the action of A on its domain.

The singular values σ_i, found on the diagonal of Σ, are the stretching factors or "gains" of the transformation along specific orthogonal directions. The largest singular value, σ_1, is precisely the *spectral norm* (or 2-norm) of A, representing the maximum factor by which A can stretch any unit vector:

$$\|A\|_2 = \max_{\|x\|=1} \|Ax\| = \sigma_1.$$

The ordered sequence $\sigma_1 \geq \sigma_2 \geq \cdots \geq \sigma_r > 0$ (where $r = \text{rank}(A)$) indicates the relative importance of different modes of the transformation.

A rapid decay in these singular values, as we shall see in Chapters 11 and 12, suggests that the matrix A (and the data it might represent) is well-approximated by a matrix of lower rank.

The columns of $V = [v_1 \ldots v_n]$ are the *right singular vectors*. Each v_i represents a *principal input direction*. When A represents data (e.g., rows as observations, columns as features), these v_i correspond to principal directions or inherent patterns within the feature space. For instance, if A were a document-term matrix, the v_i might align with underlying topics.

The columns of $U = [u_1 \ldots u_m]$ are the *left singular vectors*, forming an orthonormal basis for the output space $\mathbb{R}^m$. The fundamental action of A on its right singular vectors is to map them to scaled versions of its left singular vectors:

$$Av_i = \sigma_i u_i \quad \text{for } i = 1, \ldots, \min(m, n).$$

If $\sigma_i = 0$, then $Av_i = \mathbf{0}$. The set $\{u_1, \ldots, u_r\}$ (corresponding to non-zero σ_i) forms an orthonormal basis for the image (or column space) of A, $\text{im}(A)$. These u_i are the *principal output directions*. In a data context where rows of A are observations, the columns of $U\Sigma$ (specifically $U_r\Sigma_r$, where U_r and Σ_r contain the first r components) can be interpreted as the coordinates of the transformed data in the basis of principal output directions.

The SVD provides a striking geometric refinement and an explicit construction for the concepts surrounding the Fundamental Theorem of Linear Algebra (Theorem 6.9). Recall that any linear transformation $A : \mathbb{R}^n \to \mathbb{R}^m$ (using matrix notation for T) is associated with four fundamental subspaces. The SVD not only confirms their existence and dimensional relationships but also provides orthonormal bases for each:

- The *column space* or *image* of A, $\text{im}(A) \subset \mathbb{R}^m$: The left singular vectors $\{u_1, \ldots, u_r\}$ corresponding to non-zero singular values $\sigma_1, \ldots, \sigma_r$ form an orthonormal basis for $\text{im}(A)$. Thus, $\dim(\text{im } A) = r$.
- The *null space* or *kernel* of A, $\ker(A) \subset \mathbb{R}^n$: The right singular vectors $\{v_{r+1}, \ldots, v_n\}$ corresponding to zero singular values (if $r < n$) form an orthonormal basis for $\ker(A)$. Thus, $\dim(\ker A) = n - r$.
- The *row space* of A, $\text{im}(A^T) \subset \mathbb{R}^n$: Since $A^T = V\Sigma^T U^T$, the right singular vectors $\{v_1, \ldots, v_r\}$ (which are left singular vectors for A^T corresponding to non-zero σ_i) form an orthonormal basis for $\text{im}(A^T)$. This space is also $(\ker A)^\perp$. Thus, $\dim(\text{im } A^T) = r$.
- The *left null space* of A, $\ker(A^T) \subset \mathbb{R}^m$: The left singular vectors $\{u_{r+1}, \ldots, u_m\}$ corresponding to positions associated with zero singular values (if $r < m$) form an orthonormal basis for $\ker(A^T)$. This space is also $(\text{im } A)^\perp$. Thus, $\dim(\ker A^T) = m - r$.

The rank of A, r, is immediately identifiable as the number of non-zero singular values. The Rank-Nullity Theorem, $\dim(\ker A) + \dim(\operatorname{im} A^T) = n$ (or $(n - r) + r = n$), and its counterpart for A^T are thus made explicit by the SVD.

The SVD expresses A as a sum of r rank-one matrices, often called the SVD expansion:

$$A = U\Sigma V^T = \sum_{i=1}^{r} \sigma_i u_i v_i^T.$$

Each term $\sigma_i u_i v_i^T$ is a rank-one matrix representing an outer product. This expansion shows A as a linear combination of these fundamental rank-one "layers," ordered by the magnitude of their corresponding singular values σ_i. This form is pivotal for low-rank approximation (Chapter 12), where truncating this sum by keeping only the first k terms yields $A_k = \sum_{i=1}^{k} \sigma_i u_i v_i^T$, the best rank-$k$ approximation to A.

Finally, the SVD provides the most general and stable way to define the *pseudoinverse* $A^\dagger$ of any matrix A, extending the concept from Chapter 6. If $A = U\Sigma V^T$, its pseudoinverse is given by:

$$A^\dagger = V\Sigma^\dagger U^T.$$

Here, $\Sigma^\dagger$ is an $n \times m$ rectangular diagonal matrix. If $\Sigma_{ii} = \sigma_i > 0$, then $(\Sigma^\dagger)_{ii} = 1/\sigma_i$. If $\Sigma_{ii} = 0$, then $(\Sigma^\dagger)_{ii} = 0$. All off-diagonal entries of $\Sigma^\dagger$ are zero.

The pseudoinverse $A^\dagger$ effectively "inverts" the action of A where possible:

- It maps vectors from $\operatorname{im}(A)$ back to $\operatorname{im}(A^T)$ by "undoing" the scaling by σ_i: if $y = \sigma_i u_i \in \operatorname{im}(A)$, then $A^\dagger y = v_i$.
- It maps vectors from $(\operatorname{im} A)^\perp = \ker(A^T)$ to the zero vector in $\mathbb{R}^n$.

As discussed in Chapter 6, $A^\dagger b$ yields the minimum-norm least-squares solution to $Ax = b$. The SVD construction of $A^\dagger$ makes this general for any A. Furthermore, $AA^\dagger = U_r U_r^T$ (where $U_r = [u_1 \ldots u_r]$) is the orthogonal projection onto $\operatorname{im}(A)$, and $A^\dagger A = V_r V_r^T$ (where $V_r = [v_1 \ldots v_r]$) is the orthogonal projection onto $\operatorname{im}(A^T)$.

In essence, the SVD reveals that any linear transformation, no matter how complex its initial matrix representation, can be understood as a sequence of three fundamental geometric operations: a rotation/reflection (V^T), a scaling along axes (Σ), and another rotation/reflection (U). This profound insight is central to many modern applications of linear algebra.

10.4 Invariance & Natural Structure

The singular values emerge not merely as computational artifacts but as fundamental invariants of the transformation – quantities that remain unchanged when viewed through different orthonormal bases. If Q_1 and Q_2 are orthogonal matrices, the transformation $B = Q_1 A Q_2^T$ represents the same underlying map as A, merely viewed through different coordinates. Yet its singular values match those of A exactly, measuring intrinsic stretching factors independent of our choice of measurement frame.

Two matrix norms emerge naturally from this geometric perspective:

Definition 10.4 (Matrix Norms). For a matrix $A \in \mathbb{R}^{m \times n}$:

1. The *spectral norm* (or *2-norm*) measures maximal stretching:

$$\|A\|_2 = \max_{\|x\|=1} \|Ax\| = \sigma_1$$

2. The *Frobenius norm* measures total energy:

$$\|A\|_F = \left(\sum_{i,j} a_{ij}^2 \right)^{1/2} = \left(\sum_{i=1}^{p} \sigma_i^2 \right)^{1/2}$$

Compare: The Frobenius norm comes from the inner product in Chapter 5, Example 5.3. It provides a natural measure of approximation error, effectively counting the total squared discrepancy across all matrix entries.

These norms connect deeply to the SVD structure through the fundamental metrics $A^T A$ and AA^T in domain and codomain:

$$A^T A = V \Sigma^T \Sigma V^T = \sum_{i=1}^{p} \sigma_i^2 v_i v_i^T \quad \text{and} \quad AA^T = U \Sigma \Sigma^T U^T = \sum_{i=1}^{p} \sigma_i^2 u_i u_i^T$$

The equality of nonzero eigenvalues between these matrices now emerges naturally from singular value structure.

Even more remarkable is how singular values control composition of transformations. While eigenvalues can grow explosively under matrix multiplication, singular values satisfy delicate inequalities:

Lemma 10.5 (Singular Value Interlacing). *For matrices A and B of compatible size with singular values in descending order:*

$$\sigma_i(AB) \le \sigma_i(A)\sigma_1(B)$$

Proof Idea. The key insight comes from the *minimax principle*: the kth singular value equals the minimal spectral norm over all rank-$(k-1)$ approximations. For any k-dimensional subspace S:

$$\sigma_k(AB) = \min_{\text{rank}(X)<k} \|AB - X\|_2 \le \|A\|_2 \sigma_k(B) = \sigma_1(A)\sigma_k(B)$$

A more careful argument using the SVDs of both A and B establishes the full inequality. $\square$

This control of singular values under composition has no direct analog for eigenvalues. Even the product of two symmetric positive definite matrices can have complex eigenvalues, while their singular values remain real and well-behaved. This stability under composition helps explain why singular values often prove more useful than eigenvalues in analyzing iterative processes and error propagation.

The singular values also provide the most natural measure of matrix rank:

$$\text{rank}(A) = \#\{\sigma_i > 0\}$$

This equality illuminates the geometric meaning of rank as counting independent stretching directions. More subtly, small singular values indicate directions that are *nearly* dependent – a crucial insight for the numerical computation of rank that we shall explore in Chapter 12.

Each of these relationships – through norms, composition, and rank – reflects a different facet of how singular values capture the intrinsic character of linear transformations. Their collective power lies in uniting algebraic, geometric, and computational perspectives into a single coherent framework for understanding linear maps.

The singular values of a matrix provide more than just optimal approximation – they quantify precisely how the matrix distorts space under transformation. Recall from Chapter 1 the notion of condition number as a measure of numerical sensitivity. The SVD framework now allows us to define this concept rigorously:

Definition 10.6 (Condition Number). For a nonsingular matrix A with singular values $\sigma_1 \geq \sigma_2 \geq \cdots \geq \sigma_n > 0$, the *condition number* is the ratio

$$\text{COND}(A) = \frac{\sigma_1}{\sigma_n}$$

of largest to smallest singular values. For singular matrices, we set $\text{COND}(A) = \infty$. $\bullet$

This definition illuminates why condition number measures sensitivity. When solving $Ax = b$, the SVD shows that A stretches some directions by σ_1 while compressing others by σ_n. The ratio σ_1/σ_n thus bounds how much relative errors can be amplified when computing solutions. More precisely, small perturbations δb in the right-hand side can produce changes in solution of magnitude up to $\text{COND}(A)\|\delta b\|$.

The condition number remains unchanged under orthogonal transformations: if Q_1 and Q_2 are orthogonal, then $\text{COND}(Q_1 A Q_2) = \text{COND}(A)$. This invariance reflects that conditioning measures intrinsic sensitivity

Example: The matrix from Example 1.13 in Chapter 1,

$$A = \begin{bmatrix} 1 & 0.999 \\ 0 & 0.001 \end{bmatrix}$$

has singular values $\sigma_1 \approx 2$ and $\sigma_2 \approx 0.001$, explaining its condition number of about 2000.

rather than artifacts of particular coordinate choices. Indeed, $\text{COND}(A)$ can be characterized independently of the SVD as

$$\text{COND}(A) = \|A\|\|A^{-1}\|$$

where $\|\cdot\|$ denotes any matrix norm invariant under orthogonal transformations.

This geometric interpretation of conditioning – as measuring the eccentricity of how A transforms the unit sphere – explains its fundamental importance in numerical computation. Matrices with large condition numbers map some directions to nearly zero, making accurate recovery of inputs from outputs inherently difficult regardless of what numerical method we employ.

The singular values also illuminate the fundamental subspaces studied in Chapter 3. A transformation's kernel corresponds precisely to right singular vectors with zero singular values:

$$\ker(A) = \text{span}\{v_i : \sigma_i = 0\}$$

while its image is spanned by left singular vectors with nonzero singular values:

$$\text{im}(A) = \text{span}\{u_i : \sigma_i > 0\}$$

Think: The SVD provides an orthonormal basis for each of the four fundamental subspaces: kernel, image, cokernel, and coimage. This geometric refinement of the Fundamental Theorem reveals not just dimensions but natural coordinate systems.

This understanding transforms abstract concepts like rank and nullity into concrete geometric measurements. A singular value of zero indicates a direction that collapses under the transformation; the number of such zeros counts the nullity. The nonzero singular values measure how much each surviving direction stretches or compresses; their number gives the rank.

Example 10.7 (Matrix Completion). Consider a matrix with missing entries, like data from an incomplete survey:

$$A = \begin{bmatrix} 1 & ? & 2 \\ 2 & 1 & ? \\ ? & 2 & 3 \end{bmatrix}$$

If we suspect the true matrix has low rank (meaning many dependencies among entries), the SVD suggests a natural completion strategy: fill the missing entries to minimize the number of nonzero singular values. This geometric principle – that real data often lies near a lower-dimensional subspace – has profound implications for modern data science. ◇

While the singular values are uniquely determined (when arranged in descending order), the singular vectors exhibit constrained non-uniqueness that reflects fundamental symmetries:

- Vectors corresponding to distinct nonzero singular values are determined up to sign
- Vectors sharing a singular value can be rotated within their subspace
- Vectors corresponding to zero singular values need only form an orthonormal basis for the kernel

Example 10.8 (Image Compression). A grayscale image stored as an $m \times n$ matrix typically has full rank – every singular value is nonzero. Yet most singular values may be very small, indicating directions that contribute little to the image's visual content. Setting these small singular values to zero effectively reduces the rank while maintaining essential features. This process of low-rank approximation, to be studied in detail in Chapter 12, exemplifies how singular values guide practical computation through geometric insight. $\diamond$

This geometric understanding transforms our view of linear transformations from mere computational objects into structured entities with intrinsic character. The SVD reveals not just how to factor matrices but how to read the deepest patterns woven into linear maps themselves. These patterns – of stretching, rank, and fundamental subspaces – will guide our development of both theoretical understanding and practical algorithms in the chapters ahead.

10.5   Inner Products & the SVD

The singular value decomposition generalizes naturally to linear transformations between arbitrary inner product spaces. Given $T : V \to W$ between finite-dimensional inner product spaces, the adjoint $T^* : W \to V$ defined in Chapter 5 allows us to form $T^*T : V \to V$ and $TT^* : W \to W$. These self-adjoint operators play the role of $A^T A$ and AA^T, with their spectral properties determining the SVD structure.

Definition: For vectors $u \in W$ and $v \in V$, the tensor product $u \otimes v$ denotes the rank-one operator sending $x \mapsto \langle x, v \rangle u$. This generalizes the matrix outer product uv^T to arbitrary inner product spaces.

More precisely, let $\{v_1, \ldots, v_n\}$ be orthonormal eigenvectors of T^*T with eigenvalues $\{\sigma_1^2, \ldots, \sigma_n^2\}$. For each nonzero singular value σ_i, the vector $u_i = \frac{1}{\sigma_i} Tv_i$ is well-defined, and these vectors form an orthonormal set in W. The transformation then admits decomposition:

$$T = \sum_{i=1}^{r} \sigma_i (u_i \otimes v_i)$$

where $r = \operatorname{rank}(T)$. When expressed in orthonormal bases, this abstract decomposition yields exactly the matrix factorization $A = U\Sigma V^T$ developed earlier.

This coordinate-free perspective reveals that the SVD is not merely a matrix factorization but a fundamental property of linear transformations between finite-dimensional inner product spaces. The assumption of finite dimensionality is crucial – in infinite dimensions, the story becomes far more subtle. While compact operators between Hilbert spaces admit a similar spectral decomposition with countably many singular values approaching zero, general bounded operators may lack such decomposition entirely. This boundary between finite and infinite dimensions marks a profound transition in the structure of linear transformations.

For transformations between finite-dimensional spaces, however, the inner product structure is sufficient – without it, we cannot form adjoints or measure orthogonality, and the beautiful connection between input and output spaces through singular vectors dissolves. The inner product provides exactly the geometric structure needed for the SVD to emerge naturally, independent of any choice of coordinates.

BONUS! The spectral theory of compact operators in infinite dimensions leads to deep connections with integral equations, quantum mechanics, and functional analysis. The SVD appears there as the *Schmidt decomposition* of an integral kernel.

Example 10.9 (Finite-Dimensional Function Spaces). Consider the space $\mathcal{P}_n$ of polynomials of degree at most n, equipped with the L^2 inner product on $[0,1]$: $\langle f, g \rangle = \int_0^1 f(t)g(t)\,dt$. The differentiation operator $D : \mathcal{P}_n \to \mathcal{P}_{n-1}$ is linear, and its adjoint $D^* : \mathcal{P}_{n-1} \to \mathcal{P}_n$ involves both integration and boundary terms. Though we work with functions, the finite-dimensionality of these polynomial spaces ensures the SVD exists and is unique. ◇

Example 10.10 (Integration Operator). Consider the integration operator $T : C([0,1]) \to C([0,1])$ defined by

$$(Tf)(x) = \int_0^x f(t)\,dt$$

This operator is bounded when $C[0,1]$ is equipped with the standard inner product $\langle f, g \rangle = \int_0^1 f(t)g(t)\,dt$. Its adjoint can be found through integration by parts:

$$(T^*f)(x) = \int_x^1 f(t)\,dt$$

The composed operator T^*T then has a particularly nice form:

$$(T^*Tf)(x) = \int_0^1 \min(x,t)f(t)\,dt$$

This is an example of an integral operator with a symmetric kernel. While we cannot write down its singular values and vectors explicitly (they involve solutions to certain differential equations), we know they must form a complete orthonormal set in $C([0,1])$. This infinite sequence of singular values approaches zero, ensuring the operator is compact – a key distinction from the finite-dimensional case where singular values stay bounded away from zero unless exactly zero.

This example hints at the deeper theory of integral operators, where the SVD manifests as an infinite series rather than a finite sum. The geometric intuition remains – we are still decomposing the transformation into orthogonal components – but the infinitude of dimensions introduces subtleties of convergence & completeness absent in finite dimensions.   ◇

Latent Semantic Structure in Text

Hidden within the soil of human discourse lie the roots of text that grow into meaning. The singular value decomposition reveals these latent structures, transforming our intuitive sense that *words are known by the company they keep* into more precise mathematical insights. Through careful analysis of word co-occurrence patterns, we can uncover the deep relationships that give language its remarkable expressive power.

Consider a collection of documents represented through their word frequencies. Each document becomes a vector in a high-dimensional space where each dimension corresponds to a possible word. The complete corpus forms a term-document matrix A where entry a_{ij} represents the (weighted) occurrence of word i in document j. This matrix, though sparse and high-dimensional, contains rich structure that the SVD can expose.

The singular value decomposition $A = U\Sigma V^T$ reveals fundamental semantic patterns:

- Left singular vectors (columns of U) reveal word clusters that tend to occur together
- Right singular vectors (columns of V) identify document themes
- Singular values measure the strength of these semantic associations

More remarkably, this decomposition often captures genuine semantic relationships despite operating purely on word co-occurrence patterns. Words with similar meanings tend to appear in similar contexts, creating parallel rows in the term-document matrix. The SVD detects these parallels and groups related terms in its singular vectors.

Example 10.11 (Scientific Abstract Analysis). Consider analyzing a collection of physics abstracts. The first few left singular vectors often reveal clear semantic groupings:

1. Experimental terms: "measurement", "observation", "data", "experiment"
2. Theoretical terms: "model", "theory", "prediction", "framework"
3. Quantum terms: "state", "superposition", "entanglement", "qubit"

These groupings emerge naturally from co-occurrence patterns, without any explicit semantic knowledge provided to the algorithm.   ◇

The singular values themselves tell an interesting story. They typically follow a power law decay, with a few large values followed by many smaller ones. This suggests that most semantic content lies in a low-dimensional subspace — a phenomenon that enables both efficient document indexing and semantic search.

Several practical refinements (such as term frequency-inverse document frequency weighting) as well as more sophisticated modern methods together augment these SVD foundations. Word embeddings like *Word2Vec* create dense vector representations of words that capture subtle semantic relationships. Yet these methods still reflect the fundamental insight that meaning emerges from patterns of association — patterns that the SVD is uniquely suited to reveal.

This application exemplifies the deeper truth that linear algebra illuminates structure in seemingly unstructured data. Just as the SVD reveals preferred directions of stretching in geometric transformations, it exposes natural semantic axes in the high-dimensional space of human language. The mathematics developed in this chapter thus provides not just computational tools but genuine insight into how meaning emerges from pattern.

Historical Note: Latent Semantic Analysis (LSA), developed in the late 1980s, used the SVD as its foundational mathematical tool. This application of matrix factorization to language transformed both theoretical linguistics and practical information retrieval systems.

Sensor Networks & The Geometry of Measurement

Sensors are the nerves of the Industrial Body — temperature probes in data centers, accelerometers in smartphones, pressure gauges in industrial plants. Each device measures external aspects of reality, yet these measurements harbor systematic errors from manufacturing variations, environmental conditions, and malfunctions. The singular value decomposition provides an elegant framework for integrating and smoothing these errors through the underlying geometry of measurement.

Consider an array of n sensors measuring the same physical quantity at m different times or conditions. In an ideal world, these measurements would differ only by known physical variations. Reality proves messier — each sensor has its own gain, offset, and noise characteristics. A *measurement matrix* $M \in \mathbb{R}^{m \times n}$ contains these corrupted observations:

$$M_{ij} = g_i(s_j + \eta_{ij}) + b_i$$

where s_j is the true signal at time j, g_i and b_i are the *gain* and *bias* of sensor i, and η_{ij} represents noise.

The SVD of this measurement matrix reveals remarkable structure. After centering each sensor's readings (subtracting its mean), we obtain:

$$M = U\Sigma V^T = \sum_{k=1}^{r} \sigma_k u_k v_k^T$$

The leading singular vector v_1 often captures the true underlying signal, while subsequent singular vectors reveal systematic error patterns:

- $\sigma_1 u_1 v_1^T$ approximates the physical variation
- $\sigma_2 u_2 v_2^T$ typically shows gain variation effects
- Higher terms capture more subtle error patterns

Example 10.12 (Temperature Sensor Array). Consider a server room monitored by 100 temperature sensors sampled every minute. Over an hour of operation (60 samples), we obtain a 60×100 measurement matrix. The SVD typically reveals:

1. First singular value $\sim 10\times$ larger than second, reflecting true temperature variation
2. Second singular vector correlates with sensor positions, showing spatial bias
3. Third and beyond capture various drift and noise patterns

This decomposition enables both data cleaning and sensor fault detection.   ◇

The singular values themselves provide crucial diagnostic information. Define the *effective rank* of the measurement matrix as:

$$r_{\text{eff}} = \left(\sum_{i=1}^{r} \sigma_i^2 \right)^2 \Big/ \sum_{i=1}^{r} \sigma_i^4$$

This quantity, always between 1 and $r = \text{rank}(M)$, measures how many independent patterns exist in the data. A value near 1 suggests clean measurements dominated by the physical signal; larger values indicate significant systematic errors.

Nota bene: This effective rank formula appears in random matrix theory and quantum mechanics as the participation ratio, measuring how many components participate significantly in a system.

More sophisticated analysis uses the full SVD structure to calibrate the sensor array. If v_1 approximates the true signal direction, we can estimate each sensor's gain by comparing its response to this reference:

$$\hat{g}_i = \frac{\langle m_i, v_1 \rangle}{\|v_1\|^2}$$

where m_i is the i-th column of M (centered).

This approach to sensor calibration reveals a deeper truth about physical measurement: though raw data often appears complex and corrupted, the underlying signal typically lives in a low-dimensional subspace. Systematic errors, rather than creating pure noise, generate characteristic geometric patterns that the SVD naturally detects and isolates. Modern sensor networks extend these principles through sliding window analysis for time-varying calibration and distributed computation across large arrays, yet the core insight remains: measurement errors, seemingly complex, often possess remarkably simple structure when viewed in the right coordinates.

Nota bene: More complex calibration models can be addressed through careful analysis of the singular vectors.

Tensors & Multi-Linear Algebra

Data rarely submits to two-dimensional representation. Though matrices provide powerful tools for analyzing paired relationships, reality often demands higher-dimensional structures. A collection of RGB images varying over time; user interactions with products across different platforms and contexts; the activations of a deep neural network responding to diverse inputs through multiple layers — such data naturally organizes into multi-dimensional arrays called *tensors*. Like the transition from vectors to matrices that began our development, the leap from matrices to tensors reveals new patterns through careful extension of familiar principles.

Consider first the formal structure. A *tensor* of order k assigns a number to each choice of k indices:

$$\mathcal{T} = [t_{i_1 i_2 \cdots i_k}], \quad 1 \leq i_j \leq n_j$$

Just as matrices generalize vectors by adding a second index, tensors extend this indexing to arbitrary dimension. This algebraic definition connects naturally to the differential forms encountered in multivariable calculus — a k-form is precisely an alternating k-tensor, measuring oriented k-dimensional volumes through multi-linear maps.

The familiar operations of linear algebra extend naturally to this setting. Given a third-order tensor $\mathcal{T} \in \mathbb{R}^{n_1 \times n_2 \times n_3}$, we can extract matrices through *fibers* (fixing all but one index) or *slices* (fixing all but two indices):

$$\mathcal{T}_{:jk} = [t_{ijk}]_{i=1}^{n_1}, \quad \mathcal{T}_{i::} = [t_{ijk}]_{j,k=1}^{n_2,n_3}$$

These sections provide windows into the tensor's structure, much as row and column vectors illuminated matrices.

The extension of singular value decomposition to tensors reveals profound subtlety. While matrices admit unique decomposition into orthogonal factors, tensors resist such clean factorization. The *CP decomposition* provides one natural generalization:

$$\mathcal{T} = \sum_{r=1}^{R} \sigma_r a_r \otimes b_r \otimes c_r$$

expressing the tensor as a sum of rank-one components formed through outer products. Though elegant, this decomposition rarely achieves exact low-rank representation. The *Tucker decomposition* offers greater flexibility:

$$\mathcal{T} = \mathcal{S} \times_1 U \times_2 V \times_3 W$$

where $\mathcal{S}$ denotes a small *core tensor* and $\times_k$ represents multiplication along mode k. This structure — of a concentrated core expanded through factor matrices — echoes how SVD reveals low-dimensional structure in matrices.

Example 10.13 (Neural Network Analysis). Modern deep networks organize their learned weights as tensors, with indices spanning input channels, output channels, and spatial dimensions. A convolutional layer operating on image data uses fourth-order tensors $\mathcal{W} \in \mathbb{R}^{c_{out} \times c_{in} \times h \times w}$ to transform input feature maps. Understanding how information flows through these structures demands tensor analysis:

$$\text{output}[i,x,y] = \sum_{j,p,q} \mathcal{W}[i,j,p,q] \cdot \text{input}[j, x+p, y+q]$$

Tensor decompositions of these weights reveal learned feature hierarchies while enabling efficient computation through reduced parameterization.   ◇

Beyond raw data organization, tensors provide natural structure for modern machine learning systems. Language models process sequences through attention tensors that capture relationships across tokens, positions, and feature dimensions. Computer vision models build hierarchical representations through tensor operations that preserve spatial relationships while learning abstract features. Recommendation systems model complex interactions between users, items, and contexts through tensor factorizations that capture latent patterns.

Nota bene: The differential forms studied in calculus represent special tensors that change sign under odd permutations of their inputs. Their anti-symmetry makes them ideal for integration, just as symmetric tensors prove natural for deep learning.

Example: In a tensor of image embeddings, fibers might represent feature trajectories across similar images, while slices capture feature relationships at fixed semantic levels.

Nota bene: CP stands for CANDECOMP / PARAFAC, as you may have guessed.

Example 10.14 (Multi-modal Learning). Consider a deep learning system processing images with text descriptions. Each image-text pair generates embeddings in separate spaces, but their relationship forms a third-order tensor $\mathcal{T} \in \mathbb{R}^{n \times d_1 \times d_2}$ where:

- n indexes training examples
- d_1 represents image embedding dimension
- d_2 represents text embedding dimension

The tensor structure captures how different modalities interact — its decomposition reveals shared semantic spaces that enable cross-modal retrieval and generation. $\diamond$

The mathematics of tensors continues to evolve through modern applications. Training algorithms exploit tensor structure for efficient gradient computation. Neural architectures compress their parameters through tensor factorizations. Foundation models leverage tensor operations for processing multiple modalities simultaneously. Each development reveals new aspects of these fundamental objects, transforming abstract multi-linear maps into practical tools for artificial intelligence.

Exercises: Chapter 10

1. For the following matrices, A, compute the singular value decomposition by: (1) computing $A^T A$ and $A A^T$; (2) finding the eigenvalues and eigenvectors of each; (3) constructing the matrices $U, \Sigma,$ and V.

$$A = \begin{bmatrix} 2 & 2 \\ 1 & 3 \end{bmatrix} \quad : \quad A = \begin{bmatrix} 3 & 1 \\ 1 & 2 \\ 2 & -1 \end{bmatrix}$$

2. The *Hilbert matrix* H_n has entries $h_{ij} = \frac{1}{i+j-1}$. For example:

$$H_3 = \begin{bmatrix} 1 & 1/2 & 1/3 \\ 1/2 & 1/3 & 1/4 \\ 1/3 & 1/4 & 1/5 \end{bmatrix}$$

Compute $\text{COND}(H_3)$ using the SVD; then explain the asymptotics of $\text{COND}(H_n)$ as $n \to \infty$.

3. Show that for any matrix A: $\text{tr}(A^T A) = \sum_{i=1}^{p} \sigma_i^2$ where $p = \min\{m, n\}$.

4. For a square matrix A, prove that $|\det(A)| = \prod_{i=1}^{n} \sigma_i$. Then use this to show that A is invertible if and only if all its singular values are nonzero.

   This connects the algebraic notion of determinant with the geometric meaning of singular values as stretching factors.

5. For a nonsingular matrix A, let H be the positive square root of AA^T (why is this well-defined?). Show that $A = HQ$ for some orthogonal matrix Q.

6. IF A is invertible, how could you use the SVD to quickly compute A^{-1}? *Explain.*

7. Let $A + A^T$ and $A - A^T$ have singular values at most twice those of A. When does equality hold?

8. Prove that for any nonsingular matrix A and orthogonal matrices Q_1, Q_2:

$$\text{COND}(Q_1 A Q_2) = \text{COND}(A)$$

9. For matrices A and B of compatible size, prove that:

$$\text{COND}(AB) \leq \text{COND}(A)\text{COND}(B)$$

When does equality hold?

10. Let $x \in \mathbb{R}^n$ be a unit vector. The *Rayleigh quotient* of a matrix A is defined as $R_A(x) = x^T A^T A x$. Prove that:
    (a) $\sigma_n^2 \leq R_A(x) \leq \sigma_1^2$ for all unit vectors x
    (b) Equality holds in either bound if and only if x is the corresponding right singular vector

11. For a symmetric positive definite matrix A, prove that its singular values equal its eigenvalues. Under what conditions on a general matrix A do its singular values equal the absolute values of its eigenvalues?

12. A real matrix A is called *normal* if $AA^T = A^T A$. Are there any interesting properties that the SVD of a normal matrix exhibits?

13. Let A be an $m \times n$ matrix with $m > n$ and full column rank. Show that the condition number of $A^T A$ is the square of the condition number of A. What practical implication does this have for solving least squares problems?

14. Consider a rank-k approximation A_k to matrix A obtained by keeping only the k largest singular values. Show that:

$$\text{COND}(A_k) \leq \text{COND}(A)$$

with equality if and only if $k = \text{rank}(A)$. Explain why this means low-rank approximations tend to be better conditioned than the original matrix.

15. Let V and W be finite-dimensional inner product spaces with orthonormal bases $\{e_i\}$ and $\{f_j\}$ respectively. Given a linear transformation $T : V \to W$, its *matrix elements* are $a_{ij} = \langle Te_i, f_j \rangle$. Show that:
    (a) $\sum_{i,j} |a_{ij}|^2 = \sum_k \sigma_k^2$ where σ_k are the singular values of T
    (b) This sum is independent of the choice of orthonormal bases

16. (Challenge) Let V be a finite-dimensional inner product space and $T : V \to V$ a linear transformation. The *numerical range* of T is defined as:

$$W(T) = \{\langle Tv, v \rangle : v \in V, \|v\| = 1\}$$

Show that if T is normal (i.e., $TT^* = T^*T$), then $W(T)$ is the convex hull of the eigenvalues of T. How does this set relate to the singular values when T is not normal?

17. (Challenge) Let $A = U\Sigma V^T$ be the SVD of a matrix A. Define A_k by keeping only the first k singular values in Σ and setting the rest to zero. Prove that A_k minimizes $\|A - B\|_F$ among all matrices B of rank at most k.

This optimality property of the SVD truncation will be explored further in Chapter 12 on low-rank approximation.

Chapter 11
Principal Component Analysis

"to stretch across the heavens & step from star to star"

DEEP PATTERNS LIE HIDDEN within high-dimensional data, invisible to direct observation yet fundamental to understanding complex systems. The challenge lies not in gathering data – modern science and engineering generate observations in abundance – but in extracting meaningful structure from measurements that often span dozens or hundreds of dimensions.

Consider digital images, where each pixel intensity represents one dimension, or financial markets, where thousands of securities move in subtle correlation. Even simple physical systems, when fully instrumented, generate measurements whose dimension far exceeds human intuitive grasp. Within such spaces, important features often concentrate along a few key directions, like mineral deposits concentrated by geological processes.

Principal Component Analysis (PCA) provides the mathematical tools for uncovering these essential patterns. Through careful study of how measurements vary and correlate, PCA reveals natural coordinates aligned with the data's intrinsic structure. These coordinates – ordered by their importance in capturing variation – enable both dimension reduction and pattern discovery.

The foundations for this analysis emerged from our work with singular values in Chapter 10. There we saw how any linear transformation admits decomposition into orthogonal stretching along principal axes. PCA applies this geometric insight to data matrices, where rows represent observations and columns represent measured variables. The singular vectors of such matrices reveal natural coordinates for understanding variation, while singular values measure the strength of pattern in each direction.

Our development moves from statistical foundations through geometric intuition to practical application. Though the data we analyze may seem chaotic at first glance, careful mathematical excavation often reveals underlying simplicity. Like a sculpture emerging from rough stone through careful removal of excess material, meaningful low-dimensional structure emerges from high-dimensional data through principled dimension reduction. Our task is to develop both the theory to understand this process and the tools to effect it in practice.

11.1   Covariance & Correlation

The story of variance begins with rotation. A solid body spinning about its center of mass experiences rotational resistance determined not by total mass, but by how that mass is distributed in space. The familiar scalar moment of inertia $I = \int r^2 \, dm$ measures this resistance about a single axis, but a complete description requires the full *inertia tensor*:

$$\mathcal{I} = [\mathcal{I}_{ij}] \quad : \quad \mathcal{I}_{ij} = \int (r^2 \delta_{ij} - x_i x_j) \, dm$$

Definition: the Kronecker delta δ_{ij} evaluates to 1 if $i = j$ and 0 otherwise.

This mechanical perspective – of mass distributed about a center and its resistance to different rotations – provides surprisingly deep insight into the statistical structures we now develop.

Consider first a single random variable Z. Its *mean* or *expectation* $\mu = \mathbb{E}(Z)$ acts as a center of mass, while its *variance* $\mathbb{V}(Z) = \mathbb{E}((Z - \mu)^2)$ measures spread about this center – precisely analogous to the scalar moment of inertia of a mass distribution about its centroid. The *standard deviation* $\sigma = \sqrt{\mathbb{V}}$, like the radius of gyration in mechanics, provides a characteristic length scale of this spread.

Think: Just as a solid's resistance to rotation depends on mass distribution about its axes, a dataset's statistical structure depends on how measurements distribute about their means in different directions.

In most instances, data is discrete rather than continuous, and we can represent Z as a vector $\mathbb{Z} = (z_1, \dots, z_n)^T$. From this, we have basic statistical measures:

$$\mathbb{E}(Z) = \frac{1}{n} \sum_{i=1}^{n} z_i \quad \text{and} \quad \mathbb{V}(Z) = \frac{1}{n} \sum_{i=1}^{n} (z_i - \mathbb{E}(Z))^2$$

Nota bene: Instead of a $1/n$ in front of the variance, one often sees a $1/(n-1)$, especially in the context of statistics. This reflects the loss of one degree of freedom in estimating the mean from a sampling. For purposes of doing data science and dimension reduction, $1/n$ is the more appropriate scaling and is what we shall use throughout.

In data science and statistics, one typically *centers* the data, transforming Z to $\hat{Z} = Z - \mathbb{E}(Z)$ with mean zero. This, then, leads to a geometric interpretation of variance as $\mathbb{V}(Z) = \hat{Z}^T \hat{Z}$ with the standard deviation interpreted as dimension-scale length:

$$\sigma = \sqrt{\mathbb{V}} = \frac{1}{\sqrt{n}} \sqrt{\hat{Z}^T \hat{Z}} = \frac{1}{\sqrt{n}} \|\hat{Z}\|.$$

This interpretation is the key to understanding the geometry of data.

What happens with two random variables? Covariance and correlation are the key measures. Given random variables Y and Z, their *covariance*

$$\text{cov}(Y, Z) = \mathbb{E}(\hat{Y}\hat{Z}) = \mathbb{E}((Y - \mathbb{E}(Y))(Z - \mathbb{E}(Z)))$$
$$= \frac{1}{n}\hat{Y} \cdot \hat{Z}$$

measures their tendency to vary together. Like the off-diagonal terms of the inertia matrix, covariance captures coupling between different directions of variation. Positive covariance indicates that large values of Y tend to occur with large values of Z, while negative covariance suggests opposition – when one variable rises above its mean, the other tends to fall below.

That this is a dot product (scaled by dimension) should act as a balm to students who have suffered through a traditional statistics course. Filled with the geometric imagination that the dot product inspires, the Reader might guess what is to come.

To determine the degree of alignment (or misalignment) between two data vectors, one defines a *correlation* to be a rescaling of the covariance to lie between -1 and $+1$ with a correlation of zero connoting indepen-dence. As a formula, correlation becomes a familiar friend:

$$\text{CORR}(Y, Z) = \frac{\text{cov}(Y, Z)}{\sigma(Y)\,\sigma(Z)} = \frac{\hat{Y} \cdot \hat{Z}}{\|\hat{Y}\|\|\hat{Z}\|} = \cos\theta(\hat{Y}, \hat{Z}).$$

Truth: Correlation is not causation; but it is cosine similarity.

It is the angle between the two centered data points.

11.2   *Matrices & Data*

A single vector of data corresponds to one point in a point cloud. How then shall we proceed to work with the entire data set? The reader will by now have seen the future: a collection of data points becomes a collection of vectors, assembled into a data matrix.

For an (arbitrarily) ordered collection of d variables $Z_1, \ldots, Z_d$, center them each to $\hat{Z}_i$ and arrange them into a centered data matrix $\mathcal{X}$ with d columns. In practice, the columns of $\mathcal{X}$ correspond to individual variables and the rows correspond to samples or runs.

To estimate these quantities from data, we organize our observations into a *data matrix* $\mathcal{X} \in \mathbb{R}^{n \times d}$ where:

- Each row represents one observation;
- Each column corresponds to one variable;
- Entry x_{ij} is the jth measurement from observation i.

For instance, consider daily temperature measurements at three weather stations over one year:

$$\mathcal{X} = \begin{bmatrix} 72 & 70 & 68 \\ 75 & 74 & 71 \\ 65 & 63 & 62 \\ \vdots & \vdots & \vdots \end{bmatrix}$$

After centering by subtracting column means (analogous to shifting to center of mass coordinates in mechanics), the covariance matrix becomes:

$$[C] = \frac{1}{n}\mathcal{X}^T\mathcal{X} = \begin{bmatrix} 25.3 & 23.1 & 20.4 \\ 23.1 & 24.7 & 19.8 \\ 20.4 & 19.8 & 22.9 \end{bmatrix},$$

where we assume $\mathcal{X}$ has already been centered.

The diagonal entries show each station's temperature variance – station 1 shows slightly more variability than the others. The large positive off-diagonal terms indicate strong correlation between stations, as expected for nearby locations experiencing similar weather patterns. Yet the correlation is not perfect, with station pairs (1,2) showing stronger relationship than pairs involving station 3, suggesting it may be geographically more distant.

This covariance matrix is symmetric positive semidefinite – a property that emerges naturally from its construction. Its diagonal entries are the individual variances, while off-diagonal terms measure pairwise relationships.

Just as the inertia matrix's eigenvalues measure resistance to rotation about principal axes, the covariance matrix's eigenstructure reveals fundamental patterns of variation in our data. To better understand these patterns independent of scale, we sometimes normalize through the *correlation matrix* $[R] = [R_{ij}]$ where

$$R_{ij} = \frac{C_{ij}}{\sigma_i \sigma_j} = \frac{\mathrm{cov}(Z_i, Z_j)}{\sqrt{\mathbb{V}(Z_i)\mathbb{V}(Z_j)}}$$

This scales all entries to the interval $[-1, 1]$, measuring purely the strength and direction of linear relationships. For our temperature data, we first extract standard deviations from the diagonal entries of the covariance matrix:

$$\sigma_1 = \sqrt{25.3} \approx 5.03°, \quad \sigma_2 = \sqrt{24.7} \approx 4.97°, \quad \sigma_3 = \sqrt{22.9} \approx 4.79°$$

These measure the typical variation at each station. The complete correlation matrix then becomes:

$$[R] = \begin{bmatrix} 1.000 & 0.923 & 0.846 \\ 0.923 & 1.000 & 0.831 \\ 0.846 & 0.831 & 1.000 \end{bmatrix}$$

The covariance and correlation matrices transform abstract statistical relationships into concrete geometric objects. Their eigenvectors identify principal axes of variation, while their eigenvalues measure the strength of variation along these axes. This fusion of mechanical intuition, statistical theory, and geometric structure provides the foundation for the dimension reduction methods we shall develop. Even our simple temperature example suggests how datasets may harbor hidden simplicity: though we measured three variables, the strong correlations hint at underlying patterns waiting to be uncovered through principal component analysis.

11.3   Principal Components

The covariance matrix captures how our data varies along different directions in measurement space. Yet these directions, aligned with our original variables, may obscure simpler underlying patterns. Just as the projection operators of Chapter 6 revealed optimal approximating subspaces, we now seek coordinates aligned with the inherent structure of our data rather than arbitrary measurement choices.

Consider a centered data matrix $\mathcal{X} \in \mathbb{R}^{n \times d}$, where each row represents one observation of d variables, and each column has zero mean. The singular value decomposition studied in Chapter 10 provides exactly the transformation we seek:

Definition 11.1 (Principal Components). Given a centered data matrix $\mathcal{X}$, its *principal components* are the right singular vectors $v_1, \ldots, v_d$ from the SVD $\mathcal{X} = U\Sigma V^{T}$, ordered by decreasing singular value. Each component v_k represents a direction in the original variable space that captures maximal remaining variation after accounting for previous components. ●

These principal components transform our original variables into new features that capture the data's variation structure:

Definition 11.2 (PC Scores). Given a principal component v_k, the corresponding *principal component score* for observation $x \in \mathbb{R}^d$ is its projection $z_k = x^T v_k$ onto that direction. The scores of all observations along v_k form the kth *score vector* $z_k = \mathcal{X} v_k$. ●

Just as the orthogonal projections of Chapter 6 decomposed vectors into optimal approximating components, these score vectors decompose our data into orthogonal features of decreasing importance. The relationship between singular values and variation emerges naturally: if σ_k is the kth singular value of $\mathcal{X}$, then $\lambda_k = \sigma_k^2 / (n-1)$ gives the variance of scores along the kth principal component. These variances decrease as we move through components, reflecting how each successive direction captures maximal remaining variation in the data.

Example 11.3 (Gene Expression Data). Consider genetic expression measurements across thousands of genes in different cell types. Each row of our data matrix represents a cell, while columns record expression levels of different genes:

$$\mathcal{X} = \begin{bmatrix} \leftarrow & \text{cell 1} & \rightarrow \\ \leftarrow & \text{cell 2} & \rightarrow \\ & \vdots & \\ \leftarrow & \text{cell n} & \rightarrow \end{bmatrix} \begin{matrix} \text{gene 1} \\ \text{gene 2} \\ \vdots \\ \text{gene d} \end{matrix}$$

Though each cell's state lives in a space of dimension $d \approx 20{,}000$, biological constraints often restrict variation to a much lower-dimensional manifold. Principal component analysis reveals these constraints through directions v_k that often correspond to fundamental regulatory programs or cell state transitions.

Foreshadowing: The dimension reduction achieved through PCA previews how neural networks (Chapter 13) learn to represent high-dimensional data through lower-dimensional features.

Projecting onto the first two principal components produces a two-dimensional visualization:

$$\begin{bmatrix} z_{11} & z_{12} \\ z_{21} & z_{22} \\ \vdots & \vdots \\ z_{n1} & z_{n2} \end{bmatrix} = \mathcal{X}[v_1 \ v_2]$$

The resulting scatter plot of points (z_{i1}, z_{i2}) often reveals clusters of similar cell types or gradients of cellular differentiation – patterns invisible in the original high-dimensional space. $\diamond$

Example 11.4 (Market Returns). Daily returns of stocks in a market index provide another illuminating example. Each row of $\mathcal{X}$ represents one trading day, while columns represent different stocks. The principal components extract fundamental market factors:

- The first component v_1 typically captures market-wide movement
- Subsequent components often align with sector-specific variation

- Later components may reveal company-specific effects

The corresponding scores measure how strongly each factor influenced returns on different days, providing a natural decomposition of market behavior.

Nota bene: This sequence of directions generalizes the orthogonal bases of Chapter 4, now optimized to capture variation in data rather than arbitrary coordinate choices.

The fraction of total variance captured by the first k components provides a measure of how well they summarize our data:

$$r_k = \frac{\sum_{i=1}^{k} \lambda_i}{\sum_{i=1}^{d} \lambda_i} = \frac{\sum_{i=1}^{k} \sigma_i^2}{\sum_{i=1}^{d} \sigma_i^2}$$

When this ratio approaches 1 for small k, we have found a low-dimensional representation capturing most of the data's variation.

In practice, the choice of scaling crucially affects what patterns PCA discovers. Two standard approaches emerge:

1. *Covariance PCA*: Use centered data directly, preserving relative scales
2. *Correlation PCA*: Standardize each variable to unit variance first

The first emphasizes directions of large absolute variation; the second focuses on patterns of correlation regardless of scale. Section 11.5 will explore these choices and their implications in detail.

Principal component analysis thus provides a systematic way to replace arbitrary measurement coordinates with natural axes aligned to variation in our data. Like the optimal projections of Chapter 6 and the singular vectors of Chapter 10, these directions emerge from fundamental geometric and algebraic properties that the next section will carefully examine.

11.4   Optimality Properties

Principal components provide more than just convenient coordinates for data analysis – they emerge naturally from fundamental optimization principles. Like the orthogonal projections of Chapter 6, which minimized approximation error in geometric spaces, principal components minimize error in representing high-dimensional data through lower-dimensional summaries.

Consider first the problem of finding a single direction that best captures variation in our data. Given centered observations $\{x_1, \ldots, x_n\}$, we seek a unit vector v maximizing the variance of projections:

$$\text{maximize} \quad \frac{1}{n-1} \sum_{i=1}^{n} (x_i^T v)^2 \quad \text{subject to} \quad \|v\| = 1$$

This optimization has profound geometric meaning: we seek the direction along which our data shows greatest spread. Writing $\mathcal{X}$ for our centered data matrix, this objective becomes:

$$\frac{1}{n-1}v^T\mathcal{X}^T\mathcal{X}v = v^T[C]v$$

subject to $v^Tv = 1$.

Theorem 11.5 (Principal Component Optimality). *The first principal component v_1 maximizes $v^T[C]v$ subject to $\|v\| = 1$. Each subsequent component v_k maximizes this same objective subject to orthogonality with all previous components.*

Proof. By the Rayleigh-Ritz principle, the maximum value of $v^T[C]v$ subject to $\|v\| = 1$ equals the largest eigenvalue of $[C]$, achieved by its corresponding eigenvector. The optimization for subsequent components follows from the same principle applied to the residual variation after removing previous components. $\qquad\square$

This optimization perspective reveals PCA's fundamental character: it provides an orthogonal coordinate system optimally aligned with variation in our data. Each component extracts maximal remaining variance while maintaining orthogonality with previous components.

Example 11.6 (Image Compression). Consider a grayscale image represented as a matrix of pixel intensities. Though formally high-dimensional, natural images often concentrate variation in few directions due to spatial correlations. PCA reveals this low-dimensional structure: the first few principal components capture coherent image features, while later components often represent noise.

For instance, retaining only the first k components approximates the original image through:

$$\hat{\mathcal{X}}_k = \sum_{i=1}^{k} \sigma_i u_i v_i^T$$

where σ_i are singular values and u_i, v_i are left and right singular vectors. The fraction of variance preserved equals r_k from Section 11.3, providing a natural measure of approximation quality. $\qquad\diamond$

Beyond variance maximization, principal components possess several equivalent optimality properties:

1. They minimize mean squared reconstruction error for k-dimensional representations
2. They maximize mutual information between data and its projection (under Gaussian assumptions)

3. They provide optimal linear dimension reduction for preserving pair-wise distances

The reconstruction error perspective proves particularly illuminating. Given observations $\{x_i\}$, consider approximating each through:

$$\hat{x}_i = \sum_{j=1}^{k} z_{ij} v_j$$

where z_{ij} are scores and v_j are unit vectors. The principal components minimize:

$$\frac{1}{n} \sum_{i=1}^{n} \|x_i - \hat{x}_i\|^2$$

over all choices of k orthonormal vectors $\{v_j\}$. This optimality connects directly to the projection operators of Chapter 6: PCA provides the best rank-k approximation in terms of squared error.

Example 11.7 (Signal Denoising). Consider a time series corrupted by noise. If the underlying signal has simpler structure than the noise, principal components often achieve denoising through:

1. Project the noisy data onto principal components
2. Retain only components with variance above noise level
3. Reconstruct using these components only

The optimality of PCA ensures this procedure minimizes mean squared error under appropriate noise assumptions.   ◇

Though we have focused on statistical optimality, these same principles emerge from pure linear algebra through singular value decomposition. Chapter 12 will develop this complementary perspective, showing how PCA's optimality properties generalize to arbitrary matrix approximation problems while maintaining computational efficiency.

11.5 Preprocessing & Scaling

Real data rarely arrives in the pristine form assumed by our theoretical development. Consider monitoring an automated manufacturing process through five sensors:

$$X = \begin{bmatrix} 82.3 & 1205 & 4.2 & 7.1 & 156 \\ 85.1 & 1198 & 4.1 & 7.2 & 162 \\ 79.8 & 1210 & 4.3 & 6.8 & 145 \\ \vdots & \vdots & \vdots & \vdots & \vdots \end{bmatrix} \quad : \quad \text{cols} \Rightarrow \begin{array}{l} \text{Temperature (°C)} \\ \text{Pressure (kPa)} \\ \text{Flow (L/s)} \\ \text{pH} \\ \text{Conductivity } (\mu\text{S/cm}) \end{array}$$

Each row represents one measurement, but the variables differ dramatically in scale. Direct application of covariance PCA would be dominated entirely by pressure measurements, while potentially important variations in pH or flow rate vanish in the noise. Yet blindly standardizing each variable might discard meaningful scale information.

Example 11.8 (Scale Effects). For the manufacturing data above, the first principal component under different preprocessing choices reveals starkly different patterns:

Covariance PCA yields $v_1 \approx (0.002, 0.999, 0.001, 0.000, 0.003)^T$, essentially capturing pressure variation alone. After standardizing to unit variance, correlation PCA gives $v_1 \approx (0.51, 0.48, -0.42, 0.38, 0.44)^T$, revealing coordinated variation across all measurements. The choice fundamentally shapes what patterns we can discover.   ◇

Beyond scaling, real data suffers contamination from measurement errors, sensor failures, and genuine but extreme events. We need systematic methods to identify observations requiring special treatment. The key insight lies in measuring distance not just in terms of raw values, but in a way that accounts for the natural scales and relationships in our data:

Definition 11.9 (Mahalanobis Distance). For an observation x from a collection with mean $\bar{x}$ and covariance $[C]$, the *Mahalanobis distance* is:

$$d_M(x) = \sqrt{(x - \bar{x})^T [C]^{-1} (x - \bar{x})}$$

This metric accounts for both scale and correlation structure in measuring how far an observation deviates from typical patterns. The inverse covariance matrix $[C]^{-1}$ ensures that distances properly reflect the natural variability in each direction.   ●

Nota bene: The matrix $[C]^{-1}$ plays the role of the squared reciprocal standard deviation $1/\sigma^2$ in higher dimensions, accounting for both scale and correlation between variables.

For our engineering applications, this distance provides a natural way to identify observations that deviate markedly from typical patterns. Experience suggests that observations with Mahalanobis distances more than twice that of typical points warrant careful investigation.

Think: In one dimension, d_M reduces to distance from the mean measured in standard deviations: $d_M(x) = |x - \mu|/\sigma$. This connects to our intuition about "unusual" values in simple measurements.

Example 11.10 (Outlier Detection). Returning to our manufacturing data, most observations have Mahalanobis distances between 1.5 and 3 units. However, one measurement:

$$x = (84.2, 1203, 12.8, 7.0, 158)^T$$

yields $d_M = 4.9$, far exceeding the typical range. Though each individual measurement appears plausible, their combination suggests a process anomaly requiring investigation. The flow rate of 12.8 L/s, while not extreme in absolute terms, proves inconsistent with the observed temperature and pressure.   ◇

Missing measurements present a final challenge. When relatively rare ($< 5\%$ of entries), simply omitting affected observations suffices. More extensive missingness requires careful treatment to avoid biasing our analysis. Simple mean imputation – replacing missing values with variable averages – often distorts correlation structure. More sophisticated iterative schemes use partial PCA results to estimate missing values, though such approaches risk imposing artificial patterns.

The choice of preprocessing fundamentally shapes what patterns PCA can discover. Different choices serve different purposes:

- Covariance PCA preserves absolute scales when they carry meaning
- Correlation PCA reveals purely relational patterns
- Robust methods sacrifice efficiency for reliability with corrupted data

Sound engineering judgment, guided by careful examination of data quality and clear analysis goals, must inform these choices.

11.6 Statistical Significance

The singular value decomposition of our centered data matrix $\mathcal{X}$ provides not just optimal directions for dimension reduction but natural measures of their importance. Each singular value σ_i quantifies how strongly its corresponding direction contributes to the data's structure – precisely analogous to the dominance concepts developed in Chapter 9. The art lies in determining where to truncate this sequence, balancing fidelity of representation against simplicity of model.

Consider the sequence of partial approximations to $\mathcal{X}$ using increasing numbers of singular values:

$$\mathcal{X}_k = \sum_{i=1}^{k} \sigma_i u_i v_i^T$$

The ratio of retained to total variation provides our first measure of approximation quality:

$$r_k = \frac{\sum_{i=1}^{k} \sigma_i^2}{\sum_{i=1}^{d} \sigma_i^2}$$

This quantity, like the spectral ratio from Section 9.2, measures how thoroughly k components capture the data's essential structure.

Example 11.11 (Vibration Analysis). Consider acceleration measurements from sensors on a bridge structure, yielding singular values:

$$\sigma_1 = 12.5, \quad \sigma_2 = 9.5, \quad \sigma_3 = 3.6, \quad \sigma_4 = 1.3, \quad \sigma_5 = 1.1, \dots$$

The sharp drop after σ_2 suggests two dominant modes of vibration. Just as dominant eigenvalues governed asymptotic behavior in Chapter 9, these dominant singular values identify directions essential to the bridge's motion. The first two components capture proportion

$$r_2 = \frac{\sigma_1^2 + \sigma_2^2}{\sum_{i=1}^{d} \sigma_i^2} \approx 0.85$$

of total variation – a quantitative measure of their dominance. ◇

Nota bene: When analyzing noise-corrupted data, sharp drops in singular values often mark transition from signal to noise – a pattern explained theoretically through random matrix theory but visible even in simple examples.

This sequence of singular values connects directly to optimal approximation theory. By Section 11.4, each $\mathcal{X}_k$ provides the best rank-k approximation to $\mathcal{X}$ in terms of squared error. The truncation level k thus trades approximation accuracy against model complexity – a balance made quantitative through the singular value spectrum.

Example 11.12 (Chemical Process Data). A chemical reactor monitored through multiple sensors yields normalized singular values decreasing more gradually:

$$\sigma_1 = 2.05, \quad \sigma_2 = 1.76, \quad \sigma_3 = 1.52, \quad \sigma_4 = 1.18, \quad \sigma_5 = 0.89, \dots$$

No sharp dominance emerges, reflecting complex coupling between process variables. The cumulative proportion $r_4 \approx 0.85$ suggests retaining four components captures most significant variation while filtering sensor noise. ◇

The practical choice of how many components to retain benefits from systematic validation. By partitioning our data into training and testing sets, we can assess how well different truncation levels generalize to new observations. Components that dominate in one portion of the data should maintain their dominance in others – a principle that helps distinguish genuine structure from sampling artifacts.

Example 11.13 (Materials Spectroscopy). Consider spectroscopic measurements where each sample generates thousands of intensity readings across different wavelengths. With limited samples, distinguishing significant components becomes crucial. Cross-validation reveals that although singular values continue well past σ_{20}, predictions using more than 5-6 components often perform worse on holdout data – a sign of overfitting despite apparent dominance in training data. ◇

Sample size fundamentally affects our confidence in identified components. When analyzing n observations of d variables, singular values beyond index $\min\{n,d\}$ must be zero – a fact following directly from the SVD's matrix structure. More subtly, the ratio n/d affects stability of non-zero singular values: too few observations relative to variables can create spurious apparent structure.

These considerations transform our theoretical understanding of SVD dominance into practical guidance for data analysis. Though lacking the deterministic certainty of pure matrix algebra, they provide systematic ways to identify genuinely dominant components while guarding against over-interpretation. The true test lies not in abstract significance but in whether retained components enable better prediction, control, or understanding of the systems we study.

11.7   Beyond Linear PCA

Reality rarely submits to purely linear description. Though principal components illuminate structure within linear transformations of space, many datasets harbor intrinsic nonlinearity – their essential patterns twist and curve through measurement space like vines growing beyond their linear supports. A temperature sensor's readings may vary sinusoidally with time; chemical reaction rates couple through nonlinear dynamics; images of handwritten digits trace complex manifolds far from any linear subspace. Understanding such data requires extending our mathematical framework beyond the linear realm that has supported our development thus far.

Consider first why linear PCA might fail. When data lies near a curved surface or manifold, no linear projection can capture its true structure. The principal components, optimal though they are for linear approximation, may entirely miss the underlying simplicity. Like shadows cast by a curved wire that seem tangled and self-intersecting, linear projections can obscure rather than reveal the true form of nonlinear data.

The key insight emerging from such examples is that we must adapt our notion of dimension reduction to respect the local structure of data. Rather than seeking global linear projections, we might approximate our data locally by tangent spaces that follow its curves and bends. This perspective – that nonlinear structure often appears nearly linear when examined at sufficiently small scales – provides the foundation for modern manifold learning methods.

Example 11.14 (Single Cell Genomics). Consider gene expression measurements from cells undergoing differentiation. Though measured in a space of thousands of dimensions (one per gene), the developmental process often follows a branching path as cells progress from stem-like states toward specialized types. No linear projection captures this branching structure – PCA might show a tangled mess where biologically meaningful progression exists. The data's true simplicity emerges only when we respect its curved, branching geometry, revealing developmental trajectories that mirror known biology. ◇

One path beyond linearity leads through *kernel methods*. Rather than working directly in measurement space, we implicitly map our data into a higher-dimensional feature space through a kernel function $k(x, y)$ measuring similarity between observations. Within this expanded space, we perform standard PCA – though now operating on the kernel matrix $K = [k(x_i, x_j)]$ rather than the covariance of raw measurements.

This *kernel PCA* reveals nonlinear structure through careful choice of kernel function. The radial basis kernel

$$k(x, y) = \exp\left(-\frac{\|x - y\|^2}{2\sigma^2}\right)$$

measures local similarity, allowing the method to follow curved patterns in data. Though computationally intensive for large datasets, kernel PCA provides a mathematically elegant bridge between linear and nonlinear methods.

A different approach emerges from examining how linear PCA optimizes reconstruction error. Rather than restricting ourselves to linear mappings, we might seek general functions that compress our data while preserving its essential structure. This perspective leads naturally to *autoencoders* – neural networks trained to reconstruct their inputs through a low-dimensional bottleneck layer. Though we defer their detailed study to Chapter 13, autoencoders represent a fundamental generalization of PCA's dimension reduction principle.

Local methods provide yet another path beyond linearity. Rather than seeking global structure, techniques like *locally linear embedding* (LLE) preserve geometry by reconstructing each point from its neighbors. If x_i denotes our ith observation and $\mathcal{N}(i)$ its set of nearest neighbors, LLE first computes weights w_{ij} solving:

$$\min_{w_{ij}} \sum_{i=1}^{n} \left\| x_i - \sum_{j \in \mathcal{N}(i)} w_{ij} x_j \right\|^2 \quad \text{subject to} \quad \sum_{j \in \mathcal{N}(i)} w_{ij} = 1$$

These weights capture local geometry through linear approximations. A

Foreshadowing: The implicit feature mappings of kernel methods preview how neural networks will learn representations transforming data into more meaningful spaces.

BONUS! Beyond the methods discussed here, Topological Data Analysis (TDA) provides tools for understanding the shape of data. While beyond our current scope, TDA offers powerful insights into data geometry through linear-algebraic tools such as *persistent homology*.

second optimization then finds low-dimensional coordinates y_i preserving these relationships:

$$\min_{y_i} \sum_{i=1}^{n} \left\| y_i - \sum_{j \in \mathcal{N}(i)} w_{ij} y_j \right\|^2$$

This local-to-global principle – that complex structure often reduces to simpler pieces when properly decomposed – appears throughout mathematics and engineering. Just as integration reduces complex functions to sums of simple pieces, or finite elements approximate complex domains through simple patches, manifold learning methods build global understanding from local approximations.

Think: The two-stage optimization of LLE mirrors how manifolds appear locally linear – first capture local structure, then find global coordinates respecting these local patterns.

The emergence of these nonlinear methods marks a profound shift in how we think about dimension reduction. No longer constrained to linear projections, we can seek representations that truly capture how our data organizes itself in space. This freedom brings both power and challenge – we gain expressiveness at the cost of uniqueness and computational simplicity. The balance between these factors shapes the choice of method in practice:

- Kernel PCA offers mathematical elegance but scales poorly
- Local methods capture fine structure but may miss global patterns
- Autoencoders provide flexibility but require significant data and computation

Yet these challenges should not obscure the fundamental insight: that dimension reduction, properly understood, means finding simpler representations that preserve essential structure. Whether through kernels, local approximations, or learned transformations, the goal remains to distill complex data to its meaningful essence. This principle – that high-dimensional data often has lower-dimensional structure waiting to be discovered – will guide our development through neural networks and beyond.

The path from linear PCA to modern nonlinear methods recapitulates a broader pattern in mathematical thought. Just as classical linear algebra grew to embrace infinite-dimensional spaces and nonlinear operators, our tools for understanding data must evolve beyond the linear framework that birthed them. The methods we have explored represent first steps toward this broader vision – one that will continue to unfold as we delve deeper into neural networks and artificial intelligence in Chapter 13.

Decoding Neural Population Activity

The coordinated firing of neural populations encodes complex behaviors through patterns of electrical activity. Modern experimental techniques enable simultaneous recording from hundreds of neurons, each contributing a time-varying signal that reflects both local computation and global brain state. While individual neurons exhibit complex and often noisy dynamics, their collective activity reveals remarkably clear structure when analyzed through the lens of dimensionality reduction.

Consider a typical motor cortex experiment where researchers record from $n = 256$ neurons while a subject performs reaching movements toward different targets. Each neuron's firing rate varies with time, producing a vector $r(t) \in \mathbb{R}^n$ of instantaneous population activity sampled at millisecond resolution. Over a session with 100 reaches lasting 500ms each, we obtain the centered data matrix:

$$
R = \begin{bmatrix} \leftarrow & r(0) - \bar{r} & \rightarrow \\ \leftarrow & r(1) - \bar{r} & \rightarrow \\ & \vdots & \\ \leftarrow & r(50000) - \bar{r} & \rightarrow \end{bmatrix} \in \mathbb{R}^{50000 \times 256}
$$

where $\bar{r}$ denotes the temporal average firing rate vector. The sample covariance matrix $[C] = \frac{1}{T-1} R^T R \in \mathbb{R}^{256 \times 256}$ captures how pairs of neurons co-vary in their firing patterns.

A concrete example reveals the remarkable structure in this neural data. During a recent experiment, the first ten singular values of the normalized neural activity matrix were:

$$
\sigma_1 = 128.4, \quad \sigma_2 = 84.2, \quad \sigma_3 = 52.1, \quad \sigma_4 = 31.5, \quad \sigma_5 = 18.7
$$

$$
\sigma_6 = 12.3, \quad \sigma_7 = 8.1, \quad \sigma_8 = 5.4, \quad \sigma_9 = 3.8, \quad \sigma_{10} = 2.6
$$

The proportion of variance explained by the first k components:

$$
r_k = \frac{\sum_{i=1}^{k} \sigma_i^2}{\sum_{i=1}^{256} \sigma_i^2}
$$

reaches $r_5 = 0.86$ with just five components. This dramatic dimensionality reduction – from 256 neurons to 5 principal components while retaining 86% of variance – suggests neural computation occurs in a much lower-dimensional space than raw cell counts would indicate.

The principal components themselves, computed as eigenvectors of $[C]$, reveal distinct aspects of motor encoding. The first component v_1 shows coordinated activation across the population:

$$
v_1 = \begin{bmatrix} 0.31 \\ 0.28 \\ 0.33 \\ \vdots \\ 0.29 \end{bmatrix} \quad \text{with entries} \quad [v_1]_j \approx \frac{1}{\sqrt{n}} \pm 0.1
$$

This nearly uniform weighting suggests a global activity mode, perhaps reflecting overall movement vigor. The second component shows clear anatomical organization:

$$v_2 = \begin{bmatrix} 0.42 \\ 0.38 \\ -0.35 \\ \vdots \\ -0.41 \end{bmatrix}$$

with positive weights for neurons preferring forward reaches and negative weights for neurons preferring backward reaches.

To validate these components against behavior, we can project the neural data at each time t onto our principal components:

$$z_i(t) = v_i^T (r(t) - \bar{r})$$

The resulting scores $z_i(t)$ reveal tight coupling between neural trajectories and movement:

- $z_1(t)$ correlates strongly with movement speed ($r = 0.82$)
- $z_2(t)$ predicts reach direction ($r = 0.76$ with target angle)
- $z_3(t)$ and $z_4(t)$ capture grip force modulation

Like the manufacturing data from Section 11.4, neural recordings require careful preprocessing. Common challenges include:

1. Highly variable firing rates across neurons (some cells fire 100x more than others)
2. Non-stationary baseline activity due to brain state changes
3. Temporally correlated noise from shared inputs
4. Missing data from transiently lost neural signals

The choice between covariance and correlation-based PCA proves especially crucial. For the data above, correlation-based analysis reveals additional structure by preventing highly active neurons from dominating:

$$[\tilde{C}]_{ij} = \frac{[C]_{ij}}{\sqrt{[C]_{ii}[C]_{jj}}}$$

The correlation-based principal components highlight coordination between neurons regardless of their absolute firing rates. This often better reflects underlying computation, as neuron-specific factors like electrode placement can artificially inflate some firing rates.

Modern experiments often record from multiple brain regions simultaneously. Consider data from both primary motor cortex (M1) and dorsal premotor cortex (PMd):

$$R = \begin{bmatrix} R_{M1} \\ R_{PMd} \end{bmatrix} \quad \text{where} \quad R_{M1} \in \mathbb{R}^{50000 \times 256} \quad \text{and} \quad R_{PMd} \in \mathbb{R}^{50000 \times 128}$$

PCA on this combined data reveals both area-specific and cross-area patterns.

Some principal components load primarily on single areas:

$$v_1 = \begin{bmatrix} v_{M1} \\ 0 \end{bmatrix} \quad \text{or} \quad v_2 = \begin{bmatrix} 0 \\ v_{PMd} \end{bmatrix}$$

while others reflect coordinated computation:

$$v_3 = \begin{bmatrix} v_{M1} \\ v_{PMd} \end{bmatrix} \quad \text{with} \quad \|v_{M1}\| \approx \|v_{PMd}\|$$

The low-dimensional structure revealed by PCA directly impacts brain-computer interface (BCI) design. Rather than attempting to decode intended movement from individual neurons, modern BCIs first project neural activity onto data-driven principal components:

$$\hat{x}(t) = W \begin{bmatrix} v_1^T(r(t) - \bar{r}) \\ \vdots \\ v_k^T(r(t) - \bar{r}) \end{bmatrix}$$

where W maps the k principal component scores to decoded movement variables $\hat{x}(t)$. This approach proves remarkably robust – even if individual neurons are lost, the principal components often remain stable by leveraging the coordinated activity of the remaining population.

The statistical significance framework proves vital for validating low-dimensional neural structure. Consider splitting our reaching data into training and testing sets:

$$R_{train} = R_{1:40000,:} \quad \text{and} \quad R_{test} = R_{40001:50000,:}$$

Computing principal components on R_{train} and evaluating variance explained on R_{test} helps distinguish reliable patterns from overfitting. When neural dimensions consistently align with behavior across splits while maintaining high variance explained, we gain confidence in their computational relevance.

Beyond basic research, dimensionality reduction enables sophisticated clinical applications. A paralyzed patient using a BCI requires reliable control signals despite neural variability. PCA-based decoders maintain performance by extracting consistent movement-related dimensions even as individual neurons change their tuning. This synthesis of mathematical theory and clinical practice exemplifies how dimensionality reduction transforms our understanding of neural computation while enabling practical neural engineering.

The analysis of neural population activity demonstrates both the power and limitations of PCA. Though linear methods reveal remarkable structure, the brain's intrinsic nonlinearity and dynamics suggest even richer patterns await discovery. For now, PCA provides our clearest window into how populations of neurons collectively encode behavior – a testament to the profound utility of dimensionality reduction in modern neuroscience.

Eigenfaces

The human brain's remarkable ability to recognize faces has long inspired mathematical approaches to image analysis. Among these, the method of *eigenfaces* stands out for both its elegant simplicity and its profound influence on computer vision. Though modern deep learning approaches have superseded it for practical applications, eigenfaces provide perhaps the clearest demonstration of how PCA can extract meaningful structure from high-dimensional data.

Consider a grayscale image of size $m \times n$ pixels. Though we naturally view this as a rectangular array of intensities, we can reshape it into a single vector in $\mathbb{R}^{mn}$. A face image of modest 100×100 resolution thus becomes a point in $\mathbb{R}^{10000}$ — a space of such high dimension that direct analysis proves hopeless. Yet faces, despite their complexity, exhibit strong statistical regularities that PCA can reveal.

Given a collection of N face images $\{x_1, \ldots, x_N\}$, each reshaped into a vector, we first center the data by subtracting the mean face:

$$\bar{x} = \frac{1}{N} \sum_{i=1}^{N} x_i \quad : \quad \tilde{x}_i = x_i - \bar{x}$$

The centered images form the columns of a data matrix $\mathcal{X} \in \mathbb{R}^{(mn) \times N}$. Its singular value decomposition $\mathcal{X} = U \Sigma V^T$ yields the eigenfaces as the columns of U corresponding to the largest singular values.

These eigenfaces, when reshaped back to image format, reveal striking structure. The first few capture large-scale features — overall face shape, the position of eyes and mouth, major lighting variations. Subsequent eigenfaces encode progressively finer details, with later components often representing noise or idiosyncratic features. Typically 40-50 components suffice to capture the essential structure of face images, achieving dimension reduction by two orders of magnitude.

Any face image x can be approximated through projection onto the first k eigenfaces:

$$x \approx \bar{x} + \sum_{i=1}^{k} c_i u_i \quad \text{where} \quad c_i = (x - \bar{x})^T u_i$$

The coefficients c_i form a *face code* — a low-dimensional representation capturing the image's essential features. This encoding enables efficient face recognition: similar faces yield similar coefficients, allowing recognition through simple distance metrics in the reduced space.

The preprocessing of face images proves crucial for effective analysis. Beyond basic centering, careful attention to:

1. Alignment and scaling of facial features
2. Lighting normalization
3. Background removal
4. Handling of facial expressions

significantly impacts the quality of the resulting eigenfaces. The Mahalanobis distance in face space provides a natural way to detect images that deviate markedly from typical faces — a valuable test for reliability in practical systems.

Modern face recognition has largely moved to deep learning approaches that can handle greater variation in pose, lighting, and expression. Yet the eigenface method's core insight — that high-dimensional face images lie near a lower-dimensional manifold — remains valid. Indeed, the intermediate layers of modern neural networks often learn representations remarkably similar to eigenfaces, suggesting these patterns reflect fundamental structure rather than mere algorithmic artifacts.

Beyond its practical impact, the eigenface method exemplifies how linear algebra transforms abstract theory into engineering solutions. The singular value decomposition of Chapter 10 becomes a practical tool for dimension reduction; the preprocessing principles of Section 11.5 find concrete application; the statistical considerations of Section 11.6 guide implementation choices. This synthesis — of mathematics, statistics, and engineering — enables machines to begin approaching human capabilities in pattern recognition and scene understanding.

BONUS! The eigenface technique extends naturally to other types of image analysis. Researchers have applied similar methods to medical images (eigen-tumors for cancer detection), satellite data (eigen-landscapes for terrain classification), and industrial inspection (eigen-defects for quality control).

Exercises: Chapter 11

1. The following measurements come from two sensors monitoring a simple process:

$$\begin{bmatrix} 1.0 & 0.8 \\ 0.8 & 2.0 \\ -0.5 & -0.2 \\ -1.2 & -1.5 \end{bmatrix}$$

   Compute the sample mean and covariance matrix. Find the principal components and create a scatter plot showing both the original data points and the direction of the first principal component. What proportion of variance does this component explain?

2. A 3×3 correlation matrix has eigenvalues 2.4, 0.5, and 0.1. Without seeing the original matrix:

   (a) Find its trace and determinant
   (b) Calculate proportion of variance captured by first principal component
   (c) What can you conclude about correlations between original variables?

3. Given centered data matrix $\mathcal{X} \in \mathbb{R}^{n \times 3}$ with columns representing temperature, pressure, and concentration measurements:

   (a) Write explicit formulas for computing principal component scores
   (b) Show how to project a new measurement x onto these components
   (c) Explain why standardization might be important before this computation

4. A chemical reactor is monitored through three sensors measuring temperature (°C), pressure (kPa), and flow rate (L/s). The measurements yield covariance matrix

$$[C] = \begin{bmatrix} 16 & 8 & 0 \\ 8 & 25 & -4 \\ 0 & -4 & 9 \end{bmatrix}$$

   Find its eigenvalues and eigenvectors. What proportion of total variance is captured by the first two principal components? Give a physical interpretation of what these components might reveal about the reactor's behavior.

5. A vibration analysis system records displacement at three points on a machine, yielding covariance matrix

$$[C] = \begin{bmatrix} 4 & 2 & 1 \\ 2 & 5 & 3 \\ 1 & 3 & 6 \end{bmatrix}$$

For the measurement $x = (0.5, 2.8, -1.2)^T$:

(a) Calculate its projection onto first two principal components
(b) What proportion of the vector's energy is retained?
(c) How would standardizing the variables first change your analysis?

6. Consider a sequence of measurements arriving one at a time. Describe an efficient algorithm to:

(a) Update the sample mean incrementally
(b) Update the covariance matrix without storing all past data
(c) Periodically update principal components when needed

Why might such an incremental approach be valuable in practice?

7. In analyzing semiconductor manufacturing, three quality metrics yield correlation matrix

$$[R] = \begin{bmatrix} 1.0 & 0.8 & 0.7 \\ 0.8 & 1.0 & 0.9 \\ 0.7 & 0.9 & 1.0 \end{bmatrix}$$

Find its principal components and explain what patterns they reveal about the manufacturing process. What might explain the strong correlations between all variables?

8. A time series of strain measurements from a material under cyclic loading yields singular values:

$$\sigma_1 = 12.3, \quad \sigma_2 = 5.1, \quad \sigma_3 = 0.8, \quad \sigma_4 = 0.3, \quad \sigma_5 = 0.2$$

Based on this decay pattern, how many components would you retain? Justify your answer both mathematically and in terms of physical meaning for the material's behavior.

9. Four accelerometers mounted on a bridge structure record vertical displacement (mm) with covariance matrix

$$[C] = \begin{bmatrix} 10 & -2 & 4 & 1 \\ -2 & 8 & 0 & 3 \\ 4 & 0 & 6 & -1 \\ 1 & 3 & -1 & 5 \end{bmatrix}$$

How many principal components would you retain to capture 90% of total variance? What physical interpretation might these dominant modes have in terms of the bridge's vibration patterns?

10. Consider monitoring a chemical process where temperature ranges 150–200 °C, pressure 2000–2500 kPa, and flow 0.5–2.0 L/s. Explain why covariance PCA might be misleading here and propose a preprocessing strategy. How would you validate your choice through cross-validation?

11. In monitoring an industrial furnace, temperature measurements (°C) from five sensors yield mean vector $\bar{x} = (845, 835, 840, 850, 855)^T$ and measurement $x = (850, 823, 841, 868, 859)^T$ with covariance matrix

$$[C] = \begin{bmatrix} 100 & 85 & 82 & 75 & 70 \\ 85 & 120 & 90 & 80 & 75 \\ 82 & 90 & 110 & 85 & 80 \\ 75 & 80 & 85 & 90 & 75 \\ 70 & 75 & 80 & 75 & 95 \end{bmatrix}$$

Calculate the Mahalanobis distance of x from the mean. If typical measurements have Mahalanobis distances less than 3, does this reading warrant investigation?

12. Let $\mathcal{X}$ be a centered data matrix. Prove that principal components of the correlation matrix (obtained after standardizing each variable to unit variance) differ from those of the covariance matrix unless all original variables had equal variance. When might you prefer correlation PCA to covariance PCA in practice?

13. For the Mahalanobis distance $d_M^2(x) = (x - \bar{x})^T [C]^{-1} (x - \bar{x})$, prove it is invariant under linear transformations that preserve covariance structure. Show that its level sets are ellipsoids aligned with principal components.

14. A manufacturing quality inspection system records five measurements per part. The singular values of the centered data matrix decrease as:

$$\sigma_k = 10e^{-0.8k}, \quad k = 1, \dots, 5$$

Prove that retaining the first two components captures at least 95% of the total variance. What does the exponential decay of singular values suggest about the intrinsic dimensionality of the manufacturing variations?

15. Consider the problem of splitting data into training and validation sets before performing PCA. Explain: (a) How to properly compute principal components using only training data; (b) How to measure reconstruction error on validation data; (c) Why this cross-validation approach provides better estimates of generalization error.

16. Consider a dataset where most measurements cluster near the mean but occasional outliers appear. Explain why PCA based on the covariance matrix might be unduly influenced by these outliers. Propose a robust alternative approach using concepts from Section 11.5. How would you validate that your approach better captures the true data structure?

17. The correlation matrix for measurements in a chemical process is nearly block diagonal, with strong correlations within blocks but weak correlations between blocks. Explain what this structure reveals about the process. How might you use this information to design a hierarchical monitoring strategy based on PCA?

18. Explain how the singular value spectrum of a data matrix reveals intrinsic dimensionality of the underlying process. How might you distinguish between: (a) Genuine low-dimensional structure; (b) Random correlations from finite sampling; and (c) Systematic measurement artifacts. Support your answer using concepts from Section 11.6.

Chapter 12
Low Rank Approximation

"do I not stretch the heavens abroad or fold them up like a garment"

A MATRIX OF FULL RANK harbors redundancy in subtle forms – patterns obscured by profusion of detail, structure veiled by sheer abundance of data. The fundamental character of a matrix often lies not in its apparent complexity but in the simpler forms lurking beneath. This insight – that matrices admit approximation by versions of lower rank – transforms both our theoretical understanding and practical computation. The art lies in finding such approximations that preserve essential features while stripping away superfluity.

The singular value decomposition developed in Chapter 10 provides our first glimpse of how matrices might be approximated by simpler versions of themselves. Just as a sculptor reveals form by removing excess material, we can often capture a matrix's essential character by keeping only its dominant singular values and vectors. This process of low-rank approximation – of finding the best possible simple representation of complex data – connects deeply to both the abstract theory of matrix analysis and the practical challenges of modern computing.

Consider a matrix A whose entries represent pixel intensities in a digital image. Though formally of high rank (each entry potentially independent), the structured nature of real images often allows remarkably accurate reconstruction from far fewer parameters. The theoretical foundations developed in this chapter explain both why such compression is possible and how to achieve it optimally. These same principles guide dimension reduction in data analysis, signal processing, and the emerging field of matrix completion – where we seek to recover full matrices from partial observations.

The path ahead moves from abstract approximation theory through practical algorithms to modern applications. We begin with the fundamental limits of low-rank approximation, establishing both what is possible and what is not. This leads naturally to computational methods that achieve these limits efficiently, even for massive datasets where direct computation proves infeasible. The theory extends to handle corrupted or incomplete data, reflecting the messy reality of real-world applications. Throughout, we maintain the perspective that effective approximation requires both mathematical understanding and sound engineering judgment.

12.1   *Optimal Approximation*

The singular value decomposition developed in Chapter 10 revealed how any matrix can be expressed as a sum of rank-one terms, each capturing a different direction of variation in the data. This decomposition suggests a natural path to approximation: rather than keeping all terms, we might retain only those corresponding to the largest singular values. Such truncation raises fundamental questions about optimality – how well can matrices be approximated by simpler versions of themselves, and what form should such approximations take?

Let us begin by formalizing what we mean by rank-constrained approximation:

Definition 12.1 (Rank-k Approximation). For a matrix $A \in \mathbb{R}^{m \times n}$ and integer $k \leq \text{rank}(A)$, a *rank-k approximation* is any matrix $B \in \mathbb{R}^{m \times n}$ satisfying $\text{rank}(B) = k$. The error in such an approximation is typically measured using matrix norms introduced in Chapter 5, particularly:

1. The Frobenius norm: $\|A - B\|_F = \sqrt{\Sigma_{i,j}(a_{ij} - b_{ij})^2}$
2. The spectral norm: $\|A - B\|_2 = \sigma_1(A - B)$

where $\sigma_1(A - B)$ denotes the largest singular value of the difference matrix.

Given matrix A with singular value decomposition $A = U\Sigma V^T$, we can construct a natural rank-k approximation by keeping only the first k singular values:

$$A_k = \sum_{i=1}^{k} \sigma_i \boldsymbol{u}_i \boldsymbol{v}_i^T$$

This truncated SVD yields a matrix of rank exactly k, as each term $\sigma_i \boldsymbol{u}_i \boldsymbol{v}_i^T$ has rank one and the terms combine linearly. The error in this approximation takes a particularly elegant form:

$$\|A - A_k\|_F^2 = \sum_{i=k+1}^{r} \sigma_i^2$$

Compare: The truncation of singular values parallels the dimensionality reduction of PCA in Chapter 11, but now viewed through the lens of matrix approximation rather than statistics.

where $r = \text{rank}(A)$. Each truncated singular value contributes exactly its square to the total approximation error.

The profound importance of this construction emerges from its optimality properties:

Theorem 12.2 (Eckart-Young-Mirsky). *Let $A \in \mathbb{R}^{m \times n}$ have singular value decomposition $A = U\Sigma V^T$. Then for any matrix B of rank at most k:*

$$\|A - A_k\|_F \leq \|A - B\|_F$$

That is, the truncated SVD A_k provides the best possible rank-k approximation to A in the Frobenius norm.

Historical Note: Though named for work in the 1930s, the core insight that matrices admit optimal low-rank approximations through SVD emerged from astronomy, where Karl Pearson used it to fit planes to noisy data in 1901.

Proof. Let B have rank at most k and consider the matrix $A - B$. By the Fundamental Theorem (Theorem 3.27), its null space has dimension at least $n - k$. This space must have nontrivial intersection with the span of $\{v_{k+1}, \ldots, v_n\}$, which also has dimension $n - k$. Let x be a unit vector in this intersection.

Then $\|A - B\|_F^2 \geq \|(A - B)x\|^2 = \|Ax\|^2$ since $Bx = \mathbf{0}$. But Ax lies in $\text{span}\{u_{k+1}, \ldots, u_r\}$ and has norm at least σ_{k+1}. Therefore:

$$\|A - B\|_F^2 \geq \sigma_{k+1}^2 = \|A - A_k\|_F^2$$

establishing optimality of the truncated SVD. $\qquad\square$

This remarkable result extends beyond the Frobenius norm to any unitarily invariant matrix norm – including the spectral norm $\|A\|_2 = \sigma_1$ and the nuclear norm $\|A\|_* = \sum_i \sigma_i$ which we shall explore in Section 12.4. The geometric intuition remains: A_k captures the k most important directions of variation in the data, as measured by singular values.

Example 12.3 (Image Compression). Consider a grayscale image stored as a 1024×1024 matrix A of pixel intensities. Though formally of rank 1024, most singular values prove negligibly small. A rank-50 approximation A_{50} often captures the visually significant features while reducing storage from over 1 million values to just:

- 50 singular values $(\sigma_1, \ldots, \sigma_{50})$
- 50 left singular vectors $(u_1, \ldots, u_{50})$
- 50 right singular vectors $(v_1, \ldots, v_{50})$

totaling about 100,000 values – a 90% reduction in storage with minimal perceptual loss. $\qquad\diamond$

The truncated SVD provides more than just optimal approximation – it reveals the fundamental limits of matrix approximation. For any matrix B of rank at most k:

$$\|A - B\|_2 \geq \sigma_{k+1}(A)$$

This lower bound, achieved by A_k, quantifies precisely how well matrices of reduced rank can approximate A. The relationship extends beyond the spectral norm to other important matrix norms:

Lemma 12.4 (Approximation Bounds). *For the truncated SVD approximation* A_k:

1. $\|A - A_k\|_2 = \sigma_{k+1}$
2. $\|A - A_k\|_F^2 = \sum_{i=k+1}^{r} \sigma_i^2$
3. $\|A - A_k\|_* = \sum_{i=k+1}^{r} \sigma_i$

Moreover, these bounds are optimal: no rank-k matrix can achieve smaller error in any of these norms.

The power of these optimal approximation results emerges most clearly through contrast with constrained approximation problems. When we require additional properties – preserving certain entries exactly or maintaining non-negativity of the original matrix – closed-form solutions generally cease to exist. Yet such constrained problems remain amenable to numerical optimization through techniques we shall explore in Section 12.3.

The profound implications of optimal approximation extend far beyond mere data compression. By revealing the fundamental limits of matrix approximation, the Eckart-Young-Mirsky theorem provides theoretical foundations for:

- Principal component analysis in statistics
- Latent semantic analysis in natural language processing
- Model reduction in dynamical systems
- Collaborative filtering in recommendation systems

In each case, the existence of optimal low-rank approximations transforms seemingly complex problems into manageable computations. The art lies not in finding these approximations – the truncated SVD provides them automatically – but in choosing appropriate rank k to balance accuracy against simplicity. This choice, guided by the singular value spectrum, exemplifies the fundamental tension between model complexity and data fidelity that pervades modern data science.

12.2   *Approximation in Practice*

The elegant optimality of truncated singular values, though mathematically complete, leaves open crucial questions of implementation. When working with real data – whether from sensor arrays, image sequences, or network traffic – we must balance theoretical optimality against practical constraints of storage, computation, and reliability. These engineering concerns transform abstract approximation theory into working systems through careful consideration of how data naturally structures itself.

Consider first the storage implications. A rank-k approximation through truncated SVD of matrix $A \in \mathbb{R}^{m \times n}$ requires storing:

1. k singular values $\sigma_1, \ldots, \sigma_k$
2. k left singular vectors $u_1, \ldots, u_k \in \mathbb{R}^m$
3. k right singular vectors $v_1, \ldots, v_k \in \mathbb{R}^n$

totaling $k(m + n + 1)$ numbers. This yields compression precisely when $k(m + n + 1) < mn$. Yet this crude parameter counting ignores the crucial question: what rank suffices to capture essential behavior?

Example 12.5 (Building Sensor Network). Consider a building instrumented with 100 temperature sensors recording hourly measurements for one year. The resulting data matrix $A \in \mathbb{R}^{8760 \times 100}$ requires nominally 876,000 values to specify completely. Each entry a_{ij} gives the temperature at sensor j during hour i. Physical reality suggests strong redundancy: heat diffuses smoothly through space while following clear daily and seasonal patterns.

The singular value spectrum reveals this structure quantitatively:

$$\sigma_1 = 1247.3, \quad \sigma_2 = 423.1, \quad \sigma_3 = 89.4, \quad \sigma_4 = 12.7, \quad \sigma_5 = 3.2, \ldots$$

This rapid decay – each singular value roughly a quarter of its predecessor – confirms strong low-rank structure. The physical meaning emerges through singular vectors: u_1 captures seasonal variation, u_2 reflects daily cycles, while v_1 and v_2 reveal primary thermal zones. A rank-4 approximation yields root-mean-square error under 0.1°C while reducing storage by 95%. ◇

Different error metrics suggest different truncation strategies. The Frobenius norm discussed in Section 12.1 measures typical reconstruction error:

$$\|A - A_k\|_F = \sqrt{\sum_{i=k+1}^{r} \sigma_i^2}$$

while the spectral norm $\|A - A_k\|_2 = \sigma_{k+1}$ bounds worst-case error – crucial when approximations feed into control systems where errors might compound. For our temperature monitoring example, spectral norm bounds ensure no reconstructed value differs by more than 0.5°C from measured data when keeping six singular values.

The singular value spectrum itself provides natural indicators for approximation quality:

1. Sharp drops suggest clear truncation points, as with our temperature data where physical modes separate cleanly
2. Gradual decay without clear gaps warns that no natural low-rank approximation may exist
3. Clusters of similar singular values hint at coupled features requiring joint preservation

Example 12.6 (Image Sequence Analysis). A video stream storing 30 frames per second at 1024×1024 resolution nominally requires over 30 million values per second. Yet most motion appears smooth both spatially and temporally. By treating each second as a $1024 \times (1024 \cdot 30)$ matrix, low-rank approximation often achieves 20:1 compression while maintaining visual quality. The singular vectors naturally separate persistent background features from moving elements through their different temporal structure:

- Static background elements appear in early singular vectors with large singular values
- Moving objects emerge in later vectors with intermediate singular values
- Sensor noise concentrates in vectors with small singular values

This decomposition, reminiscent of the principal components studied in Chapter 11, reveals how matrix approximation automatically discovers meaningful structure in data. ◇

When data arrives continuously or distributes across multiple sensors, direct SVD computation may prove impossible. Such scenarios demand the streaming algorithms we shall develop in Section 12.3, building on the dominance principles studied in Chapter 9. The mathematical insight that dominant directions emerge naturally through iteration transforms theoretical optimality into practical computation.

Physical constraints often restrict allowable approximations beyond mere rank reduction:

- Non-negativity of reconstructed values
- Exact preservation of critical sensor readings
- Bounded derivatives between adjacent measurements
- Conservation laws or other physical invariants

Definition 12.7 (Constrained Low-Rank Approximation). Given matrix $A \in \mathbb{R}^{m \times n}$ and constraint set $\mathcal{C}$, the *constrained rank-k approximation problem* seeks:

$$\min_{B \in \mathcal{C}} \|A - B\|_F \quad \text{subject to} \quad \text{rank}(B) \leq k$$

Common constraints include:

1. Non-negativity: $\mathcal{C} = \{B : b_{ij} \geq 0\}$
2. Entry-wise bounds: $\mathcal{C} = \{B : l_{ij} \leq b_{ij} \leq u_{ij}\}$
3. Exact matching: $\mathcal{C} = \{B : b_{ij} = a_{ij} \text{ for } (i,j) \in \mathcal{I}\}$

where $\mathcal{I}$ denotes a set of index pairs for entries that must be preserved exactly.

Though such constrained approximations rarely admit closed forms like the Eckart-Young-Mirsky theorem of Section 12.1, the truncated SVD often provides excellent initialization for numerical optimization seeking physically valid approximations. We shall explore algorithms for solving such constrained problems in Section 12.3.

The practical power of low-rank approximation emerges precisely when theoretical structure aligns with natural data properties. Just as Chapter 11's principal components revealed patterns in high-dimensional data, truncated singular values expose redundancy in matrix measurements. This fusion of mathematical elegance with engineering reality transforms abstract optimization theory into practical tools for data compression and analysis.

12.3 Algorithms at Scale

The iterative systems studied in Chapter 9 revealed how matrix powers naturally amplify dominant directions while suppressing others. This same principle – of dominance emerging through iteration – provides our first path to efficient computation of singular vectors. Though direct SVD computation proves infeasible for large matrices, careful iteration can efficiently extract just the dominant singular vectors needed for low-rank approximation.

Definition 12.8 (Power Iteration). For matrix $A \in \mathbb{R}^{m \times n}$, the *power method* iterates:

$$x_{k+1} = \frac{A x_k}{\|A x_k\|}$$

starting from random initial vector x_0. When A has dominant singular value σ_1 strictly larger than σ_2, these iterates converge to the dominant right singular vector v_1.

Nota bene: Normalization here is not merely for convenience - it prevents numerical overflow while preserving the direction information we seek.

This convergence follows directly from singular vector expansion: writing $x_0 = \sum_i c_i v_i$ in the right singular vector basis:

$$A^k x_0 = \sum_{i=1}^{r} \sigma_i^k c_i u_i = \sigma_1^k \left(c_1 u_1 + \sum_{i=2}^{r} \left(\frac{\sigma_i}{\sigma_1} \right)^k c_i u_i \right)$$

Since $\sigma_i / \sigma_1 < 1$ for $i > 1$, repeated multiplication amplifies the dominant direction while suppressing others. The rate of convergence depends on the gap ratio σ_2 / σ_1 – a large gap between first and second singular values ensures rapid convergence.

Compare: Like the dominance concepts of Chapter 9, power iteration amplifies strongest patterns while naturally suppressing weaker ones. Nature herself works thus: greater populations grow faster, stronger signals drown out weaker ones.

For low-rank approximation, we need not just one but several singular vectors. The *block power method* addresses this by iterating with multiple vectors simultaneously:

$$X_{k+1} = \mathrm{orth}(AX_k)$$

where $\mathrm{orth}(\cdot)$ produces an orthonormal basis for the column space through QR factorization. This extracts an approximate basis for the dominant right singular subspace, though careful reorthogonalization proves crucial for numerical stability.

Theorem 12.9 (Block Power Convergence). *Let $A \in \mathbb{R}^{m \times n}$ have singular values $\sigma_1 > \sigma_2 > \cdots > \sigma_r > 0$. For block size p, the block power method converges to the span of the first p right singular vectors at rate:*

$$\| \sin \Theta(X_k, V_p) \|_2 \leq \left(\frac{\sigma_{p+1}}{\sigma_p} \right)^k$$

where $\Theta(X_k, V_p)$ denotes the principal angles between the subspace spanned by X_k and that spanned by the first p right singular vectors.

Think: The sine of principal angles measures misalignment between subspaces - like measuring how far we've strayed from true north while navigating by stars.

Krylov subspace methods achieve dramatically faster convergence by working in richer spaces. Rather than just the current iterate, they maintain the full subspace:

$$\mathcal{K}_k(A, x) = \mathrm{span}\{x, Ax, A^2 x, \ldots, A^{k-1} x\}$$

The *Lanczos process* builds an orthonormal basis for this space through a remarkably efficient three-term recurrence:

$$\beta_{j+1} q_{j+1} = A q_j - \alpha_j q_j - \beta_j q_{j-1} \tag{12.1}$$

where coefficients α_j, β_j emerge naturally during orthogonalization. This recurrence maintains both simplicity – requiring just one matrix multiplication per step – and numerical stability through careful management of orthogonality.

Historical Note: Lanczos discovered this elegant recurrence in 1950, but its power remained hidden until modern computers made large-scale computation feasible. Like a prophetic verse, its meaning emerged only with time.

The Lanczos vectors $\{q_1, \ldots, q_k\}$ provide excellent approximations to singular vectors, with errors decreasing rapidly as k grows. More remarkably, the tridiagonal matrix of coefficients:

$$T_k = \begin{bmatrix} \alpha_1 & \beta_2 & & \\ \beta_2 & \alpha_2 & \beta_3 & \\ & \beta_3 & \alpha_3 & \ddots \\ & & \ddots & \ddots \end{bmatrix}$$

captures the essential spectrum of A – its eigenvalues rapidly converge to singular values of A. This reduction from large sparse to small tridiagonal form exemplifies how careful algorithm design can extract structure efficiently.

Modern applications often involve matrices so large that even single matrix-vector products prove expensive. Randomization offers a powerful path forward through a surprisingly simple idea: project our large matrix onto a small random subspace. Consider the sketched matrix:

$$Y = A\Omega \quad \text{where} \quad \Omega \in \mathbb{R}^{n \times (k+p)}$$

has independent standard normal entries. Though apparently reckless, this random projection tends to preserve the dominant singular subspace with remarkable accuracy.

Theorem 12.10 (Random Projection). *Let $A \in \mathbb{R}^{m \times n}$ have singular value decomposition $A = U\Sigma V^T$, and let $\Omega \in \mathbb{R}^{n \times (k+p)}$ have independent standard normal entries. Then with probability at least $1 - 5e^{-p}$:*

$$\|A - Q_k Q_k^T A\|_2 \leq (1 + 11\sqrt{k/p})\sigma_{k+1}$$

where Q_k has orthonormal columns spanning the range of $Y = A\Omega$.

For streaming data where observations arrive sequentially:

$$A_t = A_{t-1} + u_t v_t^T$$

we seek to maintain a low-rank approximation without storing the full matrix. The *incremental SVD* provides an elegant solution: given rank-k approximation $A_{t-1} \approx U_{t-1}\Sigma_{t-1}V_{t-1}^T$, we can efficiently update it while maintaining fixed rank through small SVDs of matrices no larger than $(k+1) \times (k+1)$.

Theorem 12.11 (Incremental Error Growth). *The error in incremental rank-k approximation grows as:*

BONUS! The factor $(1+11\sqrt{k/p})$ reveals a profound truth: we can achieve near-optimal accuracy with random projections just slightly larger than our target rank ($p \approx k$).

Think: The challenge of streaming computation mirrors Chapter 9's iterative systems – how can we maintain essential structure while processing infinite sequences of updates?

1. *$O(\sqrt{t})$ in Frobenius norm*
2. *$O(\log t)$ in spectral norm*

where t is the number of rank-one updates processed.

This controlled degradation enables processing of essentially infinite data streams while maintaining approximate optimality. The evolution from power method through randomized algorithms to streaming updates reflects a broader pattern in computational mathematics: as problems scale beyond traditional limits, we maintain accuracy by carefully relaxing requirements – accepting probabilistic guarantees or approximate optimality in exchange for dramatic efficiency gains.

Practice suggests the following guidelines for algorithm selection:

1. Use power iteration for quick estimates of dominant direction
2. Apply block methods when several singular vectors are needed
3. Choose Lanczos for highest accuracy with moderate size
4. Employ randomized methods for extremely large problems
5. Use incremental updates for streaming data

Each method balances different tradeoffs between accuracy, efficiency, and implementation complexity.

12.4   Incompleteness & Reconstruction

Organic data frequently arrives with missing entries. Recommender systems know only a tiny fraction of user preferences; sensor networks suffer occasional failures; laboratory measurements capture only certain interactions between proteins. These scenarios share a common mathematical structure: we observe some entries of a matrix while others remain unknown, yet the underlying data often has low rank due to natural constraints or patterns.

Definition 12.12 (Matrix Completion Problem). Let $M \in \mathbb{R}^{m \times n}$ be an unknown matrix, and let $\Omega \subset \{1,\ldots,m\} \times \{1,\ldots,n\}$ denote a set of observed indices. Given observations $\{m_{ij} : (i,j) \in \Omega\}$, the *matrix completion problem* seeks to recover M under the assumption it has low rank. Formally, we aim to solve:

$$\min_{X} \operatorname{rank}(X) \quad \text{subject to} \quad x_{ij} = m_{ij} \text{ for all } (i,j) \in \Omega$$

Example: In collaborative filtering, we observe only a tiny fraction of possible user-item ratings, yet seek to predict unobserved preferences by exploiting latent low-rank structure.

This optimization problem proves challenging due to the discrete nature of matrix rank. A powerful relaxation emerges through the nuclear norm introduced in Section 12.1:

Definition 12.13 (Nuclear Norm). The *nuclear norm* of a matrix A, denoted $\|A\|_*$, equals the sum of its singular values:

$$\|A\|_* = \sum_{i=1}^{\min\{m,n\}} \sigma_i(A)$$

This norm provides a convex relaxation of matrix rank, since $\mathrm{rank}(A)$ equals the number of nonzero singular values.

If M has rank r, it admits factorization $M = UV^T$ where $U \in \mathbb{R}^{m \times r}$ and $V \in \mathbb{R}^{n \times r}$. The partial observations provide constraints:

$$\sum_{k=1}^{r} u_{ik} v_{jk} = m_{ij} \quad \text{for all } (i,j) \in \Omega$$

where u_{ik} and v_{jk} are entries of U and V respectively. Recovery becomes possible when these equations sufficiently constrain the factors.

Theorem 12.14 (Matrix Completion). *Let $M \in \mathbb{R}^{m \times n}$ be a rank-r matrix with singular value decomposition $M = U\Sigma V^T$, where $\sigma_r(M) > 0$. Let Ω contain entries sampled uniformly at random. Then with probability at least $1 - cn^{-3}$ (for some constant c), M is the unique solution to:*

$$\min_X \|X\|_* \quad \text{subject to} \quad x_{ij} = m_{ij} \text{ for all } (i,j) \in \Omega$$

provided:

1. *$|\Omega| \geq C\mu r(m+n)\log^2(m+n)$ entries are observed*
2. *The singular vectors $\{u_i\}$ and $\{v_i\}$ satisfy the* incoherence condition:

$$\max_{i,j} \left\{ \frac{m}{r}\|\Pi_U e_i\|^2, \frac{n}{r}\|\Pi_V e_j\|^2 \right\} \leq \mu$$

where Π_U and Π_V denote projection onto the column spaces of U and V respectively

Here C and μ are numerical constants independent of matrix dimensions.

The incoherence condition proves crucial – it ensures information spreads evenly through the matrix rather than concentrating in a few entries. When singular vectors align closely with coordinate axes, individual entries may become essential for recovery. The random sampling ensures we capture enough information about all directions.

Example 12.15 (Movie Recommendations). Consider a matrix where rows represent users, columns represent movies, and entries give ratings (1-5 stars). Most users rate only a tiny fraction of movies, yet patterns emerge: users with similar tastes rate films similarly; movies of similar genre receive correlated scores. These patterns manifest mathematically as low-rank structure.

For a dataset with 1000 users and 1000 movies, traditional approaches might require all million potential ratings. Yet if underlying preferences depend on just 40 latent factors (genre preferences, production values, etc.), the matrix completion theory shows we can recover accurate predictions from roughly 80,000 observed ratings – a dramatic reduction enabling practical recommender systems. ◇

Though theoretically elegant, nuclear norm minimization requires careful algorithmic treatment for large-scale problems. The *proximal gradient method* provides an efficient approach through iteration:

$$Y_k = X_k - \eta_k \nabla f(X_k)$$
$$X_{k+1} = \mathcal{S}_{\lambda \eta_k}(Y_k)$$

where $f(X)$ measures fidelity to observed entries, η_k controls step size, and $\mathcal{S}_\tau$ denotes singular value soft-thresholding:

$$\mathcal{S}_\tau(Y) = U \text{ DIAG}(\max\{\sigma_i - \tau, 0\})V^T$$

for SVD $Y = U\Sigma V^T$. This algorithm effectively performs gradient descent in the space of low-rank matrices.

More sophisticated models handle real-world complexities like noise and systematic bias. The *robust matrix completion* problem modifies our optimization to allow for errors:

$$\min_{X,S} \|X\|_* + \lambda \|S\|_1 \quad \text{subject to} \quad x_{ij} + s_{ij} = m_{ij} \text{ for } (i,j) \in \Omega$$

Here S captures sparse errors while X maintains low-rank structure. The parameter λ balances these competing goals – larger values favor low rank over error tolerance.

Example 12.16 (Sensor Networks). A network of temperature sensors distributed through a building should exhibit strong low-rank structure – nearby sensors record similar values, while daily and seasonal patterns affect all sensors systematically. When some sensors fail, matrix completion can interpolate their readings from surviving measurements. The incoherence conditions of Theorem 12.14 translate to physical insight: recovery works best when sensors spread evenly through the space, avoiding dense clusters or sparse regions. ◇

The theoretical guarantees for matrix completion reveal a profound truth: structure often renders redundant what initially appears essential. Just as Chapter 10's singular value decomposition exposed redundancy in full matrices, completion theory shows how partial observations suffice when underlying patterns exist. This principle – that structure enables reconstruction from incomplete data – appears throughout modern data science, from compressed sensing to neural network training.

12.5   Robust Matrix Factorization

Data rarely arrives pristine enough for direct approximation. Though the low-rank methods we have developed prove powerful when patterns lie hidden in high dimensions, real measurements often suffer more fundamental corruption. Sensors fail completely rather than merely adding noise; experimental procedures introduce systematic bias; malicious actors may deliberately contaminate observations. Yet in many cases, the underlying low-rank structure persists beneath these distortions, much as a crystal's fundamental symmetries remain even when its surface is marred.

Consider first how corruption manifests in matrix observations:

1. Gross errors in individual measurements
2. Systematic bias affecting entire rows or columns
3. Adversarial corruption targeting specific patterns
4. Random noise from measurement uncertainty

These distinct forms of corruption demand different mathematical treatment, yet share a common theme: they represent sparse deviations from an underlying low-rank structure.

Definition 12.17 (Robust Principal Component Analysis). The *robust principal component analysis* problem seeks to decompose an observed matrix M as:

$$M = L + S + N$$

where:

- L has low rank (capturing true patterns)
- S is sparse (representing gross errors)
- N contains small random noise

The noiseless version ($N = 0$) admits elegant theoretical analysis while retaining essential features.

This insight – that corruption often affects only a small fraction of entries while leaving most observations reliable – suggests combining the

Compare: Like the principal components of Chapter 11, robust PCA seeks fundamental patterns in data - but now explicitly modeling and separating out corruption and noise.

nuclear norm from Section 12.4 with the ℓ_1 norm to promote both low rank and sparsity:

Theorem 12.18 (Principal Component Pursuit). *Let $M = L_0 + S_0$ where L_0 has rank r and S_0 has at most k nonzero entries with random signs. Under the conditions:*

1. *$\text{rank}(L_0) \leq \rho_r \min\{m, n\}$*
2. *$\|S_0\|_0 \leq \rho_s mn$*
3. *The singular vectors of L_0 satisfy the incoherence condition of Theorem 12.14*

the solution to:

$$\min_{L,S} \|L\|_* + \lambda \|S\|_1 \quad subject\ to \quad L + S = M$$

exactly recovers L_0 and S_0 with high probability when $\lambda = 1/\sqrt{\max\{m, n\}}$ and ρ_r, ρ_s are sufficiently small constants.

BONUS! The optimization in Theorem 12.18, though convex, requires careful algorithmic treatment for large-scale problems. The *alternating direction method of multipliers* (ADMM) provides an efficient approach through iteration.

Example 12.19 (Video Surveillance). Consider a fixed camera recording a scene where most variation comes from a few moving objects against a static background. The resulting data matrix M has rows indexing frames and columns indexing pixels. The background contributes a low-rank component L (as it changes slowly if at all), while moving objects create sparse changes S. Despite occasional sensor glitches or lighting changes, robust PCA successfully separates these components:

$$\begin{bmatrix} \text{frame 1} \\ \text{frame 2} \\ \vdots \\ \text{frame } n \end{bmatrix} = \underbrace{\begin{bmatrix} \text{background} \\ \text{background} \\ \vdots \\ \text{background} \end{bmatrix}}_{\text{rank} \approx 1-4} + \underbrace{\begin{bmatrix} \text{moving objects} \\ \text{moving objects} \\ \vdots \\ \text{moving objects} \end{bmatrix}}_{\text{sparse}}$$

$\diamond$

More realistic scenarios often require additional structure in our decomposition. When corruptions display patterns – systematic bias in sensor calibration, coordinated manipulation of ratings – we might model S through specialized norms capturing this structure:

$$\min_{L,S} \|L\|_* + \lambda \Omega(S) \quad subject\ to \quad L + S = M$$

where $\Omega(\cdot)$ encodes our assumptions about corruption patterns.

Example 12.20 (Collaborative Filtering). In recommendation systems, some users deliberately manipulate ratings to promote or demote certain items. Rather than treating these as independent corruptions, we might model them through group sparsity:

$$\Omega(S) = \sum_{i=1}^{m} \sqrt{\sum_{j=1}^{n} s_{ij}^2}$$

This row-wise $\ell_{2,1}$ norm encourages entire users to be identified as manipulative rather than treating each rating independently.   ◇

When noise levels vary across observations, weighted variants prove valuable:

$$\min_{L,S} \|L\|_* + \lambda\|W \odot S\|_1 \quad \text{subject to} \quad L + S = M$$

where W contains entry-wise weights and $\odot$ denotes Hadamard (element-wise) product. This allows us to place more trust in reliable measurements while remaining suspicious of potentially corrupted ones.

Theorem 12.21 (Stable Recovery). *Under the conditions of Theorem 12.18, if $M = L_0 + S_0 + N$ where $\|N\|_F \leq \epsilon$, then the solution (L, S) to the robust PCA problem satisfies:*

$$\|L - L_0\|_F^2 + \|S - S_0\|_F^2 \leq C\epsilon^2$$

for some constant C depending only on matrix dimensions.

The theoretical guarantees for robust matrix factorization reveal a fundamental principle: structure can be recovered even from severely corrupted observations, provided the corruption itself exhibits some form of simplicity (like sparsity). This principle – that patterns become clearer when viewed through appropriate mathematical lenses – will guide our development of neural networks in Chapter 13.

The Netflix Prize

Like archaeologists reconstructing ancient texts from fragmentary scrolls, modern recommender systems face the challenge of inferring complete meaning from sparse observations. The Netflix Prize of 2006-2009 transformed this metaphorical reconstruction into precise mathematics, demonstrating how the theory of low-rank matrix approximation enables machines to predict human preference at unprecedented scale. This million-dollar challenge would unite the abstract mathematics developed in this chapter with practical machine learning, revealing deep connections between matrix factorization and collaborative filtering.

The mathematical landscape seemed clear: a massive but sparse matrix $R \in \mathbb{R}^{m \times n}$ containing roughly 100 million ratings from $m = 480,000$ users across

$n = 17,700$ movies. Each entry r_{ij} represented one user's 1-5 star rating of one movie, with over 98% of entries missing — most users rate only a tiny fraction of available films. The formal objective was simple: minimize the root mean squared error (RMSE) of predictions:

$$\text{RMSE} = \sqrt{\frac{1}{|\Omega|} \sum_{(i,j) \in \Omega} (r_{ij} - \hat{r}_{ij})^2}$$

where Ω denotes the set of observed ratings and $\hat{r}_{ij}$ represents predicted ratings.

The winning solution revealed the profound unity between theory and practice. At its core lay the singular value decomposition and its offspring — the very tools developed in Section 12.4 for matrix completion. Yet real-world recommender systems must contend with complexities our pristine theory ignores:

1. Temporal effects: viewing patterns shift over time
2. User bias: some rate generously, others harshly
3. Item bias: some films consistently rate higher than others
4. Implicit feedback: not rating a film conveys information

Pure low-rank approximation, though theoretically elegant, proved insufficient. The winning algorithm addressed these complexities through careful decomposition of the ratings matrix. For user u rating item i at time t, the predicted rating took the form:

$$\hat{r}_{ui}(t) = \mu + b_u(t) + b_i(t) + q_i^T \left(p_u + |\mathcal{N}(u)|^{-\frac{1}{2}} \sum_{j \in \mathcal{N}(u)} y_j \right)$$

Here:

- μ represents global mean rating
- $b_u(t)$ and $b_i(t)$ capture time-varying user and item biases
- Vectors $q_i, p_u \in \mathbb{R}^f$ encode f latent features
- The sum over $\mathcal{N}(u)$ incorporates implicit feedback

This decomposition embodies a profound insight: human preference, though complex, often reduces to the interaction of simpler patterns. Just as Section ?? showed how arbitrary matrices factor into rank-one components, the Netflix Prize revealed how customer preference emerges from interpretable features. The parameters were learned by solving the regularized optimization:

$$\min_{b,p,q,y} \sum_{(u,i) \in \Omega} (r_{ui} - \hat{r}_{ui})^2 + \lambda \left(\|b_u\|^2 + \|b_i\|^2 + \|p_u\|^2 + \|q_i\|^2 + \sum_{j \in \mathcal{N}(u)} \|y_j\|^2 \right)$$

matching the nuclear norm regularization developed in Section 12.4.

Implementation revealed equally important lessons about robust matrix factorization. Raw ratings proved surprisingly noisy, contaminated by:

- Varying interpretation of star ratings across users
- Temporal drift in rating patterns
- Selection bias in which movies users choose to rate
- Adversarial ratings from competing services

Historical Note: The 10% improvement target was chosen based on Netflix's internal analysis suggesting this threshold would yield meaningful improvement in user experience. Few anticipated it would take three years to achieve.

These challenges demanded the robust techniques developed in Section 12.5. By carefully modeling and removing systematic biases while remaining suspicious of extreme ratings, the winning team achieved an RMSE improvement of 10.06% — just barely crossing the threshold that had seemed impossible when the contest began. Their success demonstrated how theoretical insight enables practical engineering: without the mathematical foundation of matrix factorization, no amount of algorithmic ingenuity could have succeeded.

The legacy of the Netflix Prize extends far beyond movie recommendations. Its insights now power personalization across the digital economy, from e-commerce to music streaming to social media content ranking. Each applies the same core principle: that human preference, though seemingly chaotic, often admits low-rank approximation when viewed through appropriate mathematical lenses.

Nota bene: Though Netflix never implemented the winning algorithm due to engineering complexity, the competition's insights fundamentally transformed how industry approaches recommender systems.

•————————————————————•

Network Anomaly & Detection

Within the pulsing arteries of the internet, traffic flows between endpoints like blood through vessels, carrying the lifeblood of modern communication. Each packet's journey from source to destination leaves traces in router logs and network monitors, creating vast matrices of traffic data. Yet hidden within this digital circulatory system lie both normal patterns — the regular rhythms of email, web browsing, and video streaming — and dangerous anomalies that might signal security breaches or impending failures. The theory of low-rank approximation developed in this chapter provides powerful tools for distinguishing these patterns from background noise.

Consider a network with n nodes exchanging traffic over time. Each measurement produces a traffic matrix $X(t) \in \mathbb{R}^{n \times n}$ where entry $x_{ij}(t)$ represents the volume of data flowing from node i to node j during time interval t. Collecting these snapshots over m time periods yields a third-order tensor $\mathcal{X} \in \mathbb{R}^{n \times n \times m}$. Though seemingly complex, this traffic tensor often admits remarkably simple description through low-rank approximation.

The key insight emerges from examining the tensor's structure. Normal network traffic exhibits strong patterns:

Historical Note: Early attempts at network anomaly detection focused on simple statistical outlier tests. The application of matrix factorization techniques in the early 2000s revolutionized the field by revealing coordinated patterns invisible to simpler methods.

1. Spatial correlation: nearby nodes often share similar traffic profiles
2. Temporal periodicity: daily and weekly cycles dominate normal use
3. Low intrinsic dimension: most traffic follows predictable paths

These patterns suggest decomposing each time slice $X(t)$ into three components:

$$X(t) = L(t) + S(t) + N(t)$$

where $L(t)$ captures low-rank normal traffic patterns; $S(t)$ represents sparse anomalous flows; and $N(t)$ contains measurement noise.

This decomposition exactly matches the robust PCA framework developed in Section 12.5. By solving:

$$\min_{L,S} \|L\|_* + \lambda \|S\|_1 \quad \text{subject to} \quad L + S = X$$

for each time slice, we separate regular traffic patterns from potential anomalies.

The nuclear norm $\|L\|_*$ encourages low-rank structure in normal traffic, while the ℓ_1 norm $\|S\|_1$ promotes sparse anomalies.

Example 12.22 (DDoS Attack Detection). During a distributed denial of service (DDoS) attack, many compromised machines flood a single target with traffic. This creates a characteristic pattern in $S(t)$: many small entries in one column (multiple sources) targeting one destination. The sparsity constraint naturally highlights such coordinated anomalies even when individual flows appear innocent.

Consider a network with 1000 nodes under attack. The normal traffic component $L(t)$ typically has rank 10-20, reflecting legitimate communication patterns. During attack, $S(t)$ reveals the assault through elevated values in a single column, even though each individual source contributes only a small fraction of total traffic. $\diamond$

More sophisticated analysis exploits temporal structure through tensor decomposition. Writing the traffic tensor in matrix form by stacking time slices:

$$\mathbf{X} = \begin{bmatrix} X(1) \\ X(2) \\ \vdots \\ X(m) \end{bmatrix}$$

reveals additional patterns through its singular value spectrum. Normal traffic typically produces a sharp drop after 20-30 singular values, while anomalies create distinctive perturbations to this spectrum.

The robust decomposition framework handles several practical challenges:

- Missing data from failed monitors
- Quantization effects from packet counting
- Time-varying network topology
- Mixed normal and abnormal traffic

These issues connect directly to the matrix completion theory of Section 12.4. Just as Netflix must predict unseen ratings, network analysis must infer missing traffic measurements while remaining robust to corrupted observations.

Modern implementations extend these ideas through streaming algorithms that process traffic data in real-time. The incremental SVD techniques developed in Section 12.3 enable continuous monitoring of network health with bounded memory requirements. When anomalies appear, their projection onto the dominant singular vectors often reveals their nature:

- Port scans appear as sparse rows
- DDoS attacks create dense columns
- Worm propagation shows characteristic diagonal patterns
- Data exfiltration manifests as sustained point-to-point flows

Yet this power brings responsibility. The same techniques that detect malicious traffic could potentially be used to de-anonymize users or track communication patterns. As networks grow more central to modern life, the balance between security and privacy becomes increasingly delicate. These concerns mirror those

Nota bene: The choice of temporal resolution crucially affects what patterns emerge. Too fine a granularity amplifies noise; too coarse misses important structure. Most implementations use multiple timescales simultaneously.

we encountered with recommendation systems — mathematical tools that are powerful but not inherently ethical.

The application of low-rank approximation to network traffic analysis exemplifies a broader principle: that complex real-world systems often possess simpler underlying structure waiting to be discovered. Just as Chapter 10's singular value decomposition revealed preferred directions in abstract vector spaces, the techniques developed here expose the fundamental patterns in digital communication. As we move toward neural networks in Chapter 13, this insight — that high-dimensional data often lives near lower-dimensional manifolds — will prove increasingly valuable.

Exercises: Chapter 12

1. Let $A = \begin{bmatrix} 4 & 2 & 1 \\ 2 & 3 & 0 \\ 1 & 0 & 2 \end{bmatrix}$. Find the best rank-2 approximation A_2 using SVD truncation. Compare the Frobenius norm error $\|A - A_2\|_F$ to the theoretical minimum guaranteed by the Eckart-Young-Mirsky theorem.

2. The power method converges linearly with rate $|\sigma_2/\sigma_1|$. For matrix

$$A = \begin{bmatrix} 3 & 1 \\ 1 & 2 \end{bmatrix}$$

compute three iterations starting from $x_0 = (1,0)^T$. Estimate the convergence rate and compare to theory. What initial vector would give fastest convergence?

3. Image compression often involves partitioning into blocks before low-rank approximation. For the matrix

$$A = \begin{bmatrix} 100 & 98 & 45 & 43 \\ 97 & 95 & 47 & 44 \\ 42 & 45 & 153 & 155 \\ 44 & 46 & 152 & 154 \end{bmatrix}$$

compare the error from: (a) rank-2 approximation of the full matrix, versus (b) rank-1 approximations of each 2×2 block. Explain which approach better captures the block structure.

4. A chemical process generates noisy measurements of rank-2 structure corrupted by sparse errors:

$$M = \begin{bmatrix} 2.1 & 4.2 & -15.0 & 8.1 \\ 4.0 & 8.1 & 12.2 & 16.0 \\ 6.2 & 12.0 & 18.1 & 23.9 \\ 8.0 & 16.2 & 24.1 & 32.2 \end{bmatrix}$$

Using patterns in the data, identify likely outliers and propose a clean rank-2 structure. Justify your decomposition.

5. Consider the robust PCA problem $M = L + S$ where

$$M = \begin{bmatrix} 1 & 2 & 10 \\ 2 & 4 & 3 \\ 3 & 6 & 4 \end{bmatrix}$$

Explain why $(3,3)$ entry likely represents an outlier. Find a reasonable decomposition into low-rank L and sparse S components.

6. A rank-3 matrix $A \in \mathbb{R}^{5 \times 4}$ has first two singular values $\sigma_1 = 10$ and $\sigma_2 = 5$. Without computing A_1 explicitly, find:

   (a) The minimal achievable error $\|A - B\|_F$ for any rank-1 matrix B
   (b) The Frobenius norm difference $\|A_2 - A_1\|_F$ between best rank-2 and rank-1 approximations
   (c) The unknown singular value σ_3

7. Consider the matrix completion problem where we observe entries

$$M = \begin{bmatrix} 2 & ? & 1 \\ ? & 4 & ? \\ 1 & ? & 2 \end{bmatrix}$$

Find a rank-1 matrix consistent with these observations, or prove none exists. How many rank-2 completions exist? Relate your answer to the sampling conditions of Theorem 12.14.

8. A sensor network measures temperature at 5 locations every hour. After three days, two sensors fail simultaneously. The data matrix becomes:

$$M = \begin{bmatrix} 72 & 70 & 68 & ? & ? \\ 75 & 74 & 71 & ? & ? \\ 69 & 67 & 65 & ? & ? \\ 73 & 71 & 69 & ? & ? \\ \vdots & \vdots & \vdots & \vdots & \vdots \end{bmatrix}$$

Explain how daily and spatial patterns create low-rank structure. If the first singular value explains 85

9. Prove that for any matrix A and index k, the difference between successive best rank approximations satisfies

$$\|A_k - A_{k-1}\|_F^2 = \sigma_k^2$$

Use this to explain why examining ratios σ_k / σ_{k+1} helps choose truncation rank.

10. Let A have SVD $A = U \Sigma V^T$ and consider updating a rank-k approximation when a new column c arrives. Prove that computing $(A|c)_k$ from scratch is unnecessary - the update requires only $O(mk)$ operations where m is the row dimension. How would you maintain an approximate SVD for streaming data?

11. The randomized SVD algorithm computes $Y = AQ$ where Q has orthonormal columns. Consider the error bound:

$$\|A - QQ^T A\|_2 \leq (1 + \epsilon)\sigma_{k+1}$$

where ϵ depends on oversampling. Explain why this motivates using Q as a basis for rank-k approximation. How would you choose the oversampling parameter in practice?

12. For matrices A and B of compatible size, prove that

$$\|\Pi_{\mathrm{im}\,A} B\|_F \leq \|B\|_F$$

where $\Pi_{\mathrm{im}\,A}$ denotes orthogonal projection onto the image of A. When does equality hold? How does this relate to optimal low-rank approximation?

13. Let A be a matrix with blocks:

$$A = \begin{bmatrix} A_{11} & A_{12} \\ A_{21} & A_{22} \end{bmatrix}$$

Prove that $\operatorname{rank}(A) \geq \max\{\operatorname{rank}(A_{11}), \operatorname{rank}(A_{22})\}$. Then construct an example showing this bound can be strict. How does this inform strategies for block-wise low-rank approximation?

14. Let A be a rank-r matrix whose smallest nonzero singular value is σ_r. Prove that if $\|E\|_2 < \sigma_r/2$, then $\operatorname{rank}(A + E) \geq r$. Use this to explain why rank estimation from noisy data requires examining singular value gaps rather than just counting nonzero values.

15. Explain why matrix completion typically fails when missing entries form a regular pattern rather than occurring randomly. Construct a simple 3×3 example demonstrating this phenomenon. How does this relate to the incoherence conditions in Theorem 12.14?

16. For two rank-k matrices A and B, prove their product AB has rank at most k. Then show this bound is tight by constructing matrices achieving equality. How does this inform algorithms for approximate matrix multiplication?

17. Let $\{x_1, \ldots, x_n\}$ be unit vectors in $\mathbb{R}^m$. The Gram matrix G has entries $g_{ij} = x_i^T x_j$. Prove G has rank at most m and relate its singular values to principal angles between subspaces. How might this help analyze convergence of iterative methods?

18. The weighted low-rank approximation problem seeks to minimize:

$$\|W \odot (A - X)\|_F \quad \text{subject to} \quad \operatorname{rank}(X) \leq k$$

where $\odot$ denotes elementwise multiplication and W has nonnegative entries. Explain why the solution is no longer given by truncated SVD. Propose an iterative algorithm for this problem based on alternating minimization.

19. For an arbitrary matrix A, prove that the nuclear norm $\|A\|_*$ equals the optimal value of

$$\min_{UV^T = A} \frac{1}{2}(\|U\|_F^2 + \|V\|_F^2)$$

Use this to explain why nuclear norm minimization promotes low-rank solutions in matrix completion.

20. Consider an $n \times n$ matrix A with exponentially decaying singular values $\sigma_k = 2^{-k}$. For error tolerance ϵ:

   (a) Find the minimal rank needed for Frobenius norm approximation
   (b) Compare memory requirements of storing full versus low-rank form

(c) Analyze relative efficiency of power iteration versus randomized methods

How do your conclusions depend on matrix dimension n?

21. (Challenge) Let A be a rank-r matrix with condition number κ. Prove that after t iterations, the error in power iteration satisfies

$$\| \sin \Theta(x_t, v_1) \| \leq (1/\kappa)^t \| \tan \Theta(x_0, v_1) \|$$

where Θ denotes the angle between subspaces. What does this reveal about initialization sensitivity?

Chapter 13
Neural Networks & AI

"multitudes without number work incessant: the hewn stone is plac'd in beds of mortar mingled with the ashes of Vala"

THE RECENT SYNTHESIS OF LINEAR ALGEBRA into machine intelligence marks a profound transition both mathematical and societal. Through twelve chapters we have built a careful understanding of linear transformations – their fundamental subspaces and quotients, their geometry through singular values, their dynamics through eigentheory, their approximation through low rank structure.

Reasoning transcends linearity, requiring tools that can capture the subtle nonlinearities pervasive in natural and artificial systems alike. Neural networks provide exactly such tools, transforming our linear foundations into building blocks for artificial intelligence through careful composition and training.

Consider first why linearity alone proves insufficient. An image recognition system must somehow map pixel intensities to object categories; a language model must transform sequences of words into meaningful representations; a robot must convert sensor readings into appropriate actions. No purely linear transformation, however carefully chosen, can capture these fundamentally nonlinear relationships. Yet the mathematics we have developed remains essential – not as a complete solution, but as the foundational grammar from which neural networks construct their expressive power.

This construction proceeds through composition of simple elements. Each layer of a neural network performs a linear transformation followed by a nonlinear *activation function*. Though individually simple, these layers combine to approximate remarkably complex mappings. The low-rank approximations studied in Chapter 12 hint at this power – just as truncated singular values capture dominant patterns in data, trained networks learn to extract and transform relevant features through their layered structure. Understanding this learning process demands uniting our linear algebraic insights with optimization theory, probability, and information geometry.

Modern neural architectures have evolved far beyond simple layered networks. Convolutional networks exploit spatial structure through specialized linear operations; attention mechanisms dynamically reweigh their computations based on context; residual connections create shortcuts that ease optimization. Yet beneath these sophisticated designs lie patterns familiar from our study of linear algebra – the role of condition numbers in training stability, the importance of rank in compression, the geometry of high-dimensional representations. Even the largest language models, processing billions of parameters, build their capabilities on careful composition of fundamentally linear operations.

Our task is to bridge the gap between classical mathematics and modern artificial intelligence. We begin with the basic architecture of neural networks, showing how nonlinearity transforms simple matrix multiplication into universal function approximation. This leads naturally to gradient-based learning algorithms, where insights from matrix calculus guide efficient optimization. Modern architectures then emerge as careful refinements of these principles, each innovation grounded in mathematical understanding. Throughout, we maintain the perspective that neural networks represent not a break from classical methods but their natural evolution – the transformation of linear building blocks into learned intelligence through principled composition and training.

The quest to understand how neural networks achieve their remarkable capabilities – and what fundamental limits they might face – drives much of modern research in artificial intelligence. Though we cannot hope to resolve all mysteries, the mathematical foundations developed in this text provide essential tools for analysis and innovation. As we explore the frontiers of AI, from large language models to robot control, we shall see repeatedly how linear algebraic insights illuminate both practical engineering and deeper theoretical questions about the nature of learning and intelligence.

13.1   Beyond Linear Transformations

Reality resists reduction to purely linear description. Though the preceding chapters have built a powerful framework for understanding linear transformations – their geometry through singular values, their dynamics through eigentheory, their approximation through low rank structure

– nature demands more. The patterns that define intelligence, whether natural or artificial, trace curves and manifolds far beyond any linear subspace. Like light bending through a crystal, information must be transformed nonlinearly to reveal its deeper structure.

Consider the fundamental limitations of linearity. No purely linear transformation can separate points that are not linearly separable; no matrix multiplication alone can capture the XOR function; no sequence of linear operations can represent even simple logical decisions. Yet the brain performs such feats effortlessly, transforming raw sensory data into abstract understanding through cascades of neural activity. This suggests a profound principle: intelligence emerges not from linear operations in isolation, but from their careful composition with nonlinearity.

The mathematical foundation for this composition comes through nonlinear operations that transform their inputs componentwise:

Definition 13.1 (Activation Function). An *activation function* $\sigma : \mathbb{R} \to \mathbb{R}$ is a nonlinear function applied elementwise to vectors. Common choices include:

1. ReLU (Rectified Linear Unit): $\sigma(x) = \max\{0, x\}$
2. Sigmoid: $\sigma(x) = 1/(1 + e^{-x})$
3. Hyperbolic tangent: $\sigma(x) = \tanh(x)$

When applied to a vector x, we write $\sigma(x) = (\sigma(x_1), \ldots, \sigma(x_n))^T$.   •

These seemingly simple functions transform linear operations into powerful building blocks for computation. The ReLU function, for instance, implements a basic form of sparsity – setting negative values to zero while preserving positive ones. Though its derivative is discontinuous at zero, this very discontinuity proves crucial for efficient learning. The sigmoid and hyperbolic tangent functions provide smooth transitions between asymptotic limits, but modern practice favors ReLU for its computational simplicity and beneficial gradient properties.

Consider now how linear and nonlinear operations combine. Given input vector $x \in \mathbb{R}^n$, *weight matrix* $W \in \mathbb{R}^{m \times n}$, and *bias vector* $b \in \mathbb{R}^m$, a single neural network layer computes:

$$h = \sigma(Wx + b) \tag{13.1}$$

This composition of matrix multiplication, vector addition, and elementwise nonlinearity forms the fundamental building block of neural computation. Though each operation is simple, their combination enables remarkable expressivity:

Historical Note: The XOR function outputs 1 only when exactly one of its inputs is 1:

x_1	x_2	XOR
0	0	0
0	1	1
1	0	1
1	1	0

Minsky and Papert's 1969 book *Perceptrons* proved single-layer networks cannot learn this simple function, effectively halting neural network research for over a decade. The solution required hidden layers – a insight that would wait until the 1980s backpropagation revolution.

Historical Note: Early neural networks used sigmoid activation functions in analogy with biological neurons' firing rates. The shift to ReLU around 2011 marked a crucial advance, enabling much deeper networks through improved gradient flow.

Lemma 13.2 (Universal Approximation). *A neural network with a single hidden layer of sufficient width can approximate any continuous function on a compact domain to arbitrary precision. More precisely, given any continuous function $f : [0,1]^n \to \mathbb{R}$ and error $\epsilon > 0$, there exist weights W_1, W_2 and biases b_1, b_2 such that:*

$$\|f(x) - W_2\sigma(W_1 x + b_1) + b_2\|_\infty < \epsilon$$

for all $x \in [0,1]^n$.

This theoretical guarantee, though powerful, offers little practical guidance. Modern networks achieve their remarkable performance not through width but through depth – carefully composing many layers of alternating linear and nonlinear operations. This layered structure mirrors both biological neural networks, where different brain regions process information hierarchically, and the mathematical notion that complex functions often decompose naturally into simpler stages.

Example 13.3 (Image Recognition). Consider the task of recognizing handwritten digits. Each image arrives as a matrix of pixel intensities – a point in high-dimensional space. The variations in human handwriting create complex manifolds far from any linear subspace. A typical convolutional neural network transforms these images via stages:

1. Linear convolution detects local patterns
2. ReLU activation introduces nonlinearity
3. Pooling operations provide invariance
4. Further layers combine features hierarchically

No single linear transformation could separate these digit classes, yet their composition with nonlinearity achieves remarkable accuracy.   ◇

Keep going... For more information on CNNs, see the end of this chapter.

The power of neural networks lies not in the complexity of their individual operations – both the linear transformations and activation functions remain simple – but in how these operations combine to approximate complex functions. Like the crystalline structure of minerals emerging from simple atomic arrangements, intelligence emerges from careful composition of fundamental mathematical operations.

This perspective transforms our understanding of what neural networks compute. Rather than viewing them as black boxes, we see them as learnable compositions of well-understood mathematical operations. The linear transformations studied in previous chapters provide the framework for information flow, while activation functions introduce the nonlinearity essential for complex computation. This unity of linear and nonlinear operations – of continuous transformation and discrete decision – provides the foundation for modern artificial intelligence.

The sections ahead will explore how these basic building blocks combine into sophisticated architectures, how networks learn through gradient-based optimization, and how modern innovations like attention mechanisms and residual connections extend these fundamental principles. Throughout, we maintain the perspective that neural networks represent not a break from classical mathematics but its natural evolution – the transformation of linear algebraic operations into learned intelligence through principled composition and training.

13.2   Network Architecture & Matrix Factorization

The layered composition of linear and nonlinear operations discussed in Section 13.1 suggests a natural architecture for neural computation. Like the singular value decomposition of Chapter 10, neural networks factorize complex transformations into simpler components – not through analytically derived singular vectors, but through learned weight matrices separated by nonlinearity. This perspective transforms neural networks from biological metaphor into mathematical entity, revealing deep connections between classical matrix analysis and modern machine learning.

Consider a neural network with L layers. Each layer performs the transformation described by equation (13.1), but now we chain these operations together:

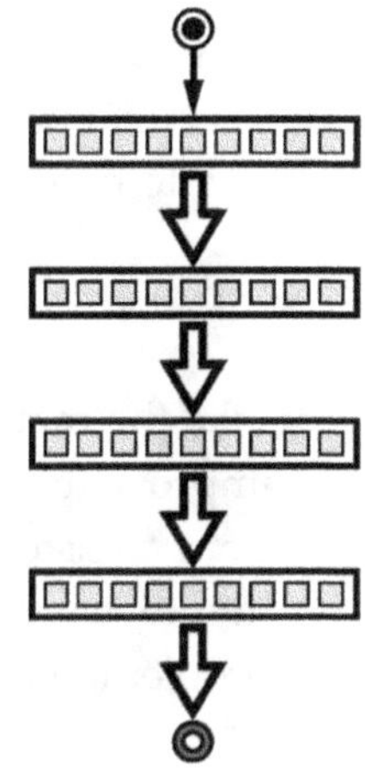

$$\begin{aligned}
h_1 &= \sigma(W_1 x + b_1) \\
h_2 &= \sigma(W_2 h_1 + b_2) \\
&\vdots \\
y &= W_L h_{L-1} + b_L
\end{aligned} \qquad (13.2)$$

where x denotes input, h_i are *hidden layers*, and y provides output. Without the nonlinear functions σ, this would reduce to mere matrix multiplication $W_L \cdots W_1$ – precisely the form we studied in the context of matrix factorization. The activation functions transform this linear composition into something far more expressive, yet the underlying structure of matrix multiplication remains essential.

Definition 13.4 (Feedforward Neural Network). A *feedforward neural network* is a function $f : \mathbb{R}^n \to \mathbb{R}^m$ parameterized by weight matrices $\{W_i\}$ and bias vectors $\{b_i\}$ that transforms its input through successive application of equation (13.2). The *width* of layer i equals the number of rows in W_i, while the *depth* equals the total number of layers L.

The choice between network depth and width presents fundamental trade-offs analogous to those encountered in low-rank approximation. Just as truncating different numbers of singular values balances approximation accuracy against complexity, the width and depth of neural networks control their expressive power. Wide but shallow networks effectively learn a large basis expansion of their target function – like keeping many singular values in a low-rank approximation. Deep networks instead compose simpler transformations, potentially capturing hierarchical structure more efficiently:

Theorem 13.5 (Deep Network Expressivity). *There exist functions that can be computed by deep neural networks with polynomial size but require exponential size when restricted to bounded depth. More precisely, for any n there exists a function $f : \{0,1\}^n \to \{0,1\}$ computable by a network of depth $O(n)$ and total size $O(n)$ that requires size $2^{n/\log n}$ when restricted to depth $O(\log n)$.*

Modern architectures enhance this basic structure through innovations that echo ideas from matrix conditioning. *Skip connections* allow information to bypass layers directly:

$$h_{i+1} = \sigma(W_i h_i + b_i) + h_i$$

Like the regularization techniques of Chapter 12, these residual paths improve numerical stability by providing direct routes for gradient flow. They transform the learning task from approximating the desired function to learning its refinements – often a better-conditioned optimization problem. Similarly, *normalization layers* standardize their inputs:

$$\hat{h}_i = \frac{h_i - \mu_i}{\sqrt{\sigma_i^2 + \epsilon}}$$

where μ_i, σ_i estimate mean and standard deviation. This operation controls the condition numbers of weight matrices by stabilizing the scale of intermediate representations, much as the column scaling studied in Chapter 1 improves matrix conditioning.

These architectural choices reflect a profound principle: the effectiveness of neural networks emerges not just from their expressive power, but from how their structure enables stable optimization. Like the carefully chosen basis transformations that revealed structure in classical linear algebra, modern network architectures create pathways for efficient learning through their thoughtful composition of operations. The next section will explore how this learning process operates through systematic application of matrix calculus.

13.3 Chains & Backpropagation

The challenge of neural network training lies not in understanding what to optimize – clearly we seek parameters that minimize prediction error – but in computing how small changes in parameters affect network output. With potentially billions of parameters spread across many layers, direct calculation of derivatives seems hopelessly complex. Yet the chain rule of multivariable calculus provides exactly the tool we need, transforming an apparently intractable computation into an elegant recursive algorithm.

Consider first the mathematical structure of what we wish to compute. Given a network with parameters Ψ (encoding all weights and biases), we seek to minimize some loss function $\mathcal{L}(\Psi)$ measuring prediction error on our training data. The challenge lies in computing $[\partial\mathcal{L}/\partial\Psi]$, the derivatives of loss with respect to parameters. Though $\mathcal{L}$ is ultimately scalar-valued, it emerges from complex composition of many operations.

Common choices for the loss function include:

- Mean squared error: $\mathcal{L}(y,\hat{y}) = \frac{1}{n}\sum_{i=1}^{n}(y_i - \hat{y}_i)^2$
- Cross-entropy for classification: $\mathcal{L}(y,\hat{y}) = -\sum_{i=1}^{n} y_i \log(\hat{y}_i)$
- L1 loss for robustness: $\mathcal{L}(y,\hat{y}) = \frac{1}{n}\sum_{i=1}^{n} |y_i - \hat{y}_i|$

where y denotes true values and $\hat{y}$ the network's predictions.

Each layer of the network performs a simple transformation of its inputs. For a layer with weights $W_\ell \in \mathbb{R}^{n_\ell \times n_{\ell-1}}$ and biases $b_\ell \in \mathbb{R}^{n_\ell}$, we compute:

$$h_\ell = \sigma(W_\ell h_{\ell-1} + b_\ell)$$

where σ is a nonlinear activation function applied componentwise, and $h_{\ell-1} \in \mathbb{R}^{n_{\ell-1}}$ denotes the previous layer's activations. The derivatives of this operation with respect to parameters are straightforward:

$$\left[\frac{\partial}{\partial W_\ell}(W_\ell h_{\ell-1} + b_\ell)\right] = \begin{bmatrix} h_{\ell-1}^T & 0 \end{bmatrix} \quad \text{and} \quad \left[\frac{\partial}{\partial b_\ell}(W_\ell h_{\ell-1} + b_\ell)\right] = I$$

The challenge comes in combining these simple pieces through the chain rule.

Definition 13.6 (Error Signal). The *error signal* $[\delta_\ell]$ at layer ℓ is the derivative of loss with respect to that layer's pre-activation output:

$$[\delta_\ell] = \left[\frac{\partial\mathcal{L}}{\partial z_\ell}\right] \in \mathbb{R}^{1 \times n_\ell}$$

where $z_\ell = W_\ell h_{\ell-1} + b_\ell \in \mathbb{R}^{n_\ell}$ denotes the layer's pre-activation values. •

Historical Note: Backpropagation first appeared in the 1974 thesis of Werbos and was subsequently ignored. Its importance for neural networks was only recognized in the 1980s through parallel discoveries by multiple researchers. The key insight – applying the chain rule systematically through recursive computation – comes from reverse-mode automatic differentiation techniques from the 1960s.

Nota bene: The terms *loss function* and *cost function* are used interchangeably in the literature, with "loss" being more common in modern machine learning practice.

Lemma 13.7 (Backpropagation Rule). *Let $\mathcal{L}$ be a scalar loss function of network output. For any layer ℓ in a neural network:*

$$[\delta_\ell] = [\delta_{\ell+1}]\, W_{\ell+1}\, \mathrm{DIAG}(\sigma'(z_\ell))$$

where σ' denotes the derivative of the activation function. The parameter gradients are then:

$$\left[\frac{\partial \mathcal{L}}{\partial W_\ell}\right] = [\delta_\ell]^T\, h_{\ell-1}^T \quad and \quad \left[\frac{\partial \mathcal{L}}{\partial b_\ell}\right] = [\delta_\ell]^T$$

Proof. By the chain rule:

$$\left[\frac{\partial \mathcal{L}}{\partial z_\ell}\right] = \left[\frac{\partial \mathcal{L}}{\partial z_{\ell+1}}\right]\left[\frac{\partial z_{\ell+1}}{\partial h_\ell}\right]\left[\frac{\partial h_\ell}{\partial z_\ell}\right]$$

The result follows from computing each factor: $[\partial z_{\ell+1}/\partial h_\ell] = W_{\ell+1}$ and $[\partial h_\ell/\partial z_\ell] = \mathrm{DIAG}(\sigma'(z_\ell))$, noting carefully the dimensions of each matrix product. $\square$

Example 13.8 (Simple Neural Network). Consider a network with two hidden layers of width n_1 and n_2 processing input $x \in \mathbb{R}^{n_0}$:

$$\begin{aligned} h_1 &= \sigma(W_1 x + b_1) \\ h_2 &= \sigma(W_2 h_1 + b_2) \\ y &= W_3 h_2 + b_3 \end{aligned}$$

where $W_1 \in \mathbb{R}^{n_1 \times n_0}$, $W_2 \in \mathbb{R}^{n_2 \times n_1}$, and $W_3 \in \mathbb{R}^{n_3 \times n_2}$. For mean squared error loss $\mathcal{L} = \frac{1}{2}\|y - t\|^2$ with target t, backpropagation computes:

$$\begin{aligned} [\delta_3] &= [y - t]^T \\ [\delta_2] &= [\delta_3]\, W_3\, \mathrm{DIAG}(\sigma'(z_2)) \\ [\delta_1] &= [\delta_2]\, W_2\, \mathrm{DIAG}(\sigma'(z_1)) \end{aligned}$$

These error signals then yield parameter gradients through matrix multiplication with stored activations. $\diamond$

The algorithm's efficiency emerges from careful organization of computation:

- Forward pass stores activations h_ℓ and pre-activations z_ℓ
- Backward pass computes error signals $[\delta_\ell]$ from output to input
- Parameter gradients emerge from simple matrix operations using stored values

This recursive structure connecting forward and backward passes echoes the complementary subspaces first encountered in Chapter 3. Just as kernel and image decomposed linear transformations, the forward and backward passes of backpropagation decompose the flow of information through the network. The error signals δ_ℓ measure precisely how changes propagate backward through this structure.

Theorem 13.9 (Backpropagation Complexity). *For a network with Λ layers each of width at most n, backpropagation computes all parameter derivatives in time $O(\Lambda n^2)$ using storage $O(\Lambda n)$.*

This remarkable efficiency – linear in depth and quadratic in width – transforms neural network training from theoretical possibility to practical reality. Like the matrix factorizations studied in Chapter 10, backpropagation achieves its power through careful decomposition of computation. The chain rule provides the mathematical foundation, while thoughtful algorithm design turns this insight into efficient implementation.

The fundamental insight of backpropagation – that influence flows backward through the network just as information flows forward – provides more than mere computational efficiency. It reveals how neural networks learn through systematic measurement of parameter influence, guided by precise mathematical principles. This fusion of calculus with computational graphs transforms the abstract possibility of gradient-based learning into practical reality.

Foreshadowing: The efficiency of backpropagation becomes crucial when we scale to deep networks with billions of parameters. The next section will show how stochastic sampling further reduces computational cost.

13.4   Stochastic Gradient Descent

The mathematics of scale prompts probabilistic thinking. Though backpropagation provides an elegant algorithm for computing gradients, applying it to every training example proves prohibitively expensive for large datasets. Like a river finding efficient paths through complex terrain, stochastic approximation transforms exact but intractable computations into efficient estimates through careful sampling. This principle – that randomness can accelerate computation while maintaining accuracy – pervades modern machine learning.

Definition 13.10 (Stochastic Gradient Descent). Let $\{\Psi_t\}_{t\geq 0}$ denote a sequence of parameter vectors updated iteratively according to:

$$\Psi_{t+1} = \Psi_t - \eta_t \left[\widehat{\frac{\partial \mathcal{L}}{\partial \Psi}}\right]^T$$

where $\eta_t > 0$ is the learning rate and $[\widehat{\partial \mathcal{L}/\partial \Psi}]$ denotes a stochastic estimate of the loss gradient.

Consider the true gradient of our loss function $\mathcal{L}$ averaged over a dataset of n examples:

$$\left[\frac{\partial \mathcal{L}}{\partial \Psi}\right] = \frac{1}{n} \sum_{i=1}^{n} \left[\frac{\partial \mathcal{L}_i}{\partial \Psi}\right]$$

Computing this exactly requires a full pass through the dataset, yet the sum's structure suggests a natural approximation. By sampling a mini-batch $\mathcal{B}$ of size $b \ll n$ uniformly at random, we obtain an unbiased estimate:

$$\left[\frac{\widehat{\partial \mathcal{L}}}{\partial \Psi}\right] = \frac{1}{b} \sum_{i \in \mathcal{B}} \left[\frac{\partial \mathcal{L}_i}{\partial \Psi}\right]$$

This stochastic gradient provides the foundation for modern neural network training.

Lemma 13.11 (Mini-batch Properties). *Let σ^2 denote the variance of individual gradient estimates. The mini-batch gradient estimator satisfies:*

1. Unbiasedness: $\mathbb{E}\left[\dfrac{\widehat{\partial \mathcal{L}}}{\partial \Psi}\right] = \left[\dfrac{\partial \mathcal{L}}{\partial \Psi}\right]$

2. Variance bound: $\mathbb{V}\left[\dfrac{\widehat{\partial \mathcal{L}}}{\partial \Psi}\right] = \mathbb{E}\left\|\left[\dfrac{\widehat{\partial \mathcal{L}}}{\partial \Psi}\right] - \left[\dfrac{\partial \mathcal{L}}{\partial \Psi}\right]\right\|_F^2 \leq \dfrac{\sigma^2}{b}$

where $\| \cdot \|_F$ denotes the Frobenius norm.

Historical Note: The term "gradient descent" arose from early focus on scalar-valued functions whose derivatives naturally formed gradient vectors. Modern usage encompasses matrix derivatives, though the name persists.

Example 13.12 (Binary Classification). Consider training a simple network to classify points in $\mathbb{R}^2$ using the logistic loss. Given parameters $w \in \mathbb{R}^2$ and $c \in \mathbb{R}$, the network computes:

$$p(x) = \frac{1}{1 + e^{-(w^T x + c)}}$$

For a dataset of points $\{(x_i, y_i)\}_{i=1}^{n}$ with $y_i \in \{0, 1\}$, the loss becomes:

$$\mathcal{L}(w, c) = -\frac{1}{n} \sum_{i=1}^{n} [y_i \log p(x_i) + (1 - y_i) \log(1 - p(x_i))]$$

With mini-batch size $b = 2$, each iteration:

1. Samples two points i, j uniformly at random
2. Computes predictions $p(x_i), p(x_j)$
3. Updates parameters using averaged gradient:

$$\begin{aligned}
w_{t+1} &= w_t - \tfrac{1}{2}\eta_t \sum_{k \in \{i,j\}} (p(x_k) - y_k) x_k \\
c_{t+1} &= c_t - \tfrac{1}{2}\eta_t \sum_{k \in \{i,j\}} (p(x_k) - y_k)
\end{aligned}$$

The stochastic updates prove far more efficient than full-batch computation while maintaining convergence. $\diamond$

Like the low-rank approximations studied in Chapter 12, mini-batch derivatives trade accuracy for computational efficiency. The connection to matrix conditioning emerges through momentum methods, which transform first-order updates into an approximation of second-order information:

$$
\begin{aligned}
m_t &= \beta m_{t-1} + (1 - \beta) \left[\frac{\partial \mathcal{L}}{\partial \Psi} \right]^T \\
\Psi_{t+1} &= \Psi_t - \eta_t m_t
\end{aligned}
$$

where $\beta \in [0,1)$ controls momentum and $m_0 = 0$. This averaging of derivatives reduces variance while accelerating convergence, particularly when optimization surfaces resemble the elongated valleys characteristic of ill-conditioned systems.

Nota bene: Like the iterative methods of Chapter 9, momentum accumulates information across steps to accelerate convergence. The parameter $\beta = 0.9$ proves effective in many applications.

In order to prove results on the convergence of SGD, a few more technical terms will need to be deployed. These are not essential to our story, but are included for precision.

Definition 13.13 (Smoothness and Strong Convexity). A differentiable function $f : \mathbb{R}^n \to \mathbb{R}$ is:

1. *L-smooth* if its gradient is Lipschitz continuous with parameter $L > 0$:

$$
\|\nabla f(x) - \nabla f(y)\| \leq L\|x - y\| \quad \text{for all } x, y \in \mathbb{R}^n
$$

Ouch! Yeah, you might need more math if you want to learn AI for real

2. *μ-strongly convex* if for some $\mu > 0$:

$$
f(y) \geq f(x) + \nabla f(x)^T (y - x) + \frac{\mu}{2} \|y - x\|^2 \quad \text{for all } x, y \in \mathbb{R}^n
$$

In effect, L-smoothness provides an upper bound on how quickly the gradient can change, while μ-strong convexity ensures a minimum amount of curvature in all directions.

Nota bene: These conditions relate directly to matrix conditioning: for quadratic functions $f(x) = \frac{1}{2} x^T A x$, the ratio L/μ equals the condition number of A.

Theorem 13.14 (SGD Convergence). *Let $\mathcal{L}$ be μ-strongly convex and L-smooth. For learning rate schedule $\eta_t = \frac{2}{\mu(t+1)}$ and mini-batch size b, stochastic gradient descent converges in expectation:*

$$
\mathbb{E}[\|\Psi_t - \Psi^*\|^2] \leq \frac{4L}{b\mu^2(t+1)} \max\{\|\Psi_0 - \Psi^*\|^2, \sigma^2\} = O\left(\frac{1}{t}\right)
$$

where Ψ^* denotes the optimal parameters and σ^2 bounds the variance of individual gradients. For non-convex losses typical in deep learning, convergence to local minima occurs under suitable regularity conditions.

Interpretation: The error decreases at least as fast as $1/t$, with the rate controlled by the problem's condition number L/μ and the gradient noise level σ^2/b.

Modern practice suggests several refinements of these basic principles:

1. Gradient clipping to bound update magnitudes:
2. Learning rate schedules that decrease over time:

3. Layer-wise adaptive rates for deep networks

4. Regularization through stochastic noise addition

Each refinement addresses specific challenges in optimization while maintaining the core idea of stochastic approximation.

The remarkable effectiveness of stochastic descent emerges from deep mathematical principles. Like the least squares methods of Chapter 6, which approximate optimal solutions through tractable computation, mini-batch derivatives provide suitable directions for optimization. The noise inherent in these estimates often proves beneficial, helping networks escape poor local minima while providing implicit regularization. This fusion of stochastic approximation with classical optimization transforms neural network training from theoretical possibility to practical reality.

Historical Note: The development of adaptive methods like Adam and RMSprop in the 2010s transformed deep learning by making training more robust to hyperparameter choices, effectively learning separate learning rates for each parameter.

13.5 *Attention & Transformers*

The synthesis of linear algebra into intelligence reaches its fullest expression in the attention mechanism of transformer architectures. Like the divine smith measuring ethereal rays with compass and scale, attention introduces a profound innovation: matrices whose entries themselves emerge dynamically from data. Each element becomes not merely a fixed weight but a learned relationship – a measure of relevance computed through systematic inner products. This mechanism – elegant in conception yet powerful in application – transforms the fixed weights studied throughout this text into adaptive transformations that focus computation where needed.

The intuition behind attention emerges naturally from information retrieval systems. Consider a library where each book has both descriptive metadata (title, subjects, keywords) and actual content. A user's search query must somehow match against the metadata to determine which books' content to examine and combine. This process – of computing relevance scores between a query and many possible keys, then using these scores to combine values – provides the template for neural attention.

The mathematics begins with three types of learned projections that transform input vectors into representations serving distinct roles. Given input vectors $\{x_1, \ldots, x_n\}$, we first project each into a query vector representing what that position is looking for, a key vector encoding what it offers to others, and a value vector containing its actual content. These roles emerge through learned matrices W_Q, W_K, and W_V respectively:

$$Q = XW_Q, \quad K = XW_K, \quad V = XW_V$$

where X stacks the input vectors as rows. Like the coordinate transfor-

Think: The query-key-value paradigm generalizes how we naturally organize information. A query expresses what we seek, keys provide searchable descriptors, while values hold actual content to be retrieved.

mations studied in Chapter 4, these projections align input features with natural axes for the attention computation – but now these axes themselves emerge through learning.

Definition 13.15 (Attention Mechanism). Given input matrix $X \in \mathbb{R}^{n \times d}$, the *attention mechanism* computes output Y through three stages:

$$A = QK^T, \quad \tilde{A}_{ij} = \frac{\exp(a_{ij}/\sqrt{d_k})}{\sum_{j=1}^{n} \exp(a_{ij}/\sqrt{d_k})}, \quad Y = \tilde{A}V \qquad (13.3)$$

where d_k denotes the dimension of query/key space. The exponential normalization ensures positive weights summing to one while sharpening focus on strong matches.

Historical Note: The attention mechanism emerged from neural machine translation in 2015, but its true power became apparent only with the transformer architecture in 2017.

Consider how this mechanism processes natural language. In the line "Tyger, tyger, burning bright," each word's representation passes through the attention computation. The query projection learns to capture what context each word needs – "burning" might query for objects it modifies. The key projection learns what context each word provides – "tyger" develops a strong key signal as the subject. The value projection learns to encode meaning that can be usefully combined – perhaps capturing that "bright" modifies the burning quality. The attention weights then determine how these meanings flow between positions.

The remarkable effectiveness of this mechanism emerges from its fusion of classical and adaptive computation. Though built from familiar operations – matrix multiplication and normalization – their careful composition creates a transformation that adapts to its input. The core matrix product QK^T measures relevance between all pairs of positions through their inner products, while the normalization ensures these scores form proper attention distributions.

Theorem 13.16 (Attention Properties). *The attention mechanism satisfies:*

$$\text{row-stochastic:} \quad \sum_{j=1}^{n} \tilde{a}_{ij} = 1 \quad \text{and} \quad \tilde{a}_{ij} \geq 0$$

$$\text{permutation equivariant:} \quad \pi(Y) = \text{Attention}(\pi(X))$$

for any permutation π. The computation requires $O(n^2 \max\{d_k, d_v\})$ operations and $O(n^2)$ memory.

Rather than compute a single attention pattern, transformer architectures typically employ multiple attention "heads" operating in parallel.

Each head learns different query/key/value projections, allowing the network to capture multiple types of relationship simultaneously. A learned output transformation W_O then combines these parallel streams:

$$\text{MultiHead}(X) = W_O \begin{bmatrix} \text{head}_1 \\ \vdots \\ \text{head}_h \end{bmatrix}$$

Think: Like the parallel paths in residual networks or the multiple singular vectors of SVD, multiple attention heads allow the network to capture different aspects of relationship simultaneously.

The complete transformer architecture synthesizes these attention blocks with several innovations that echo concepts from earlier chapters. Layer normalization stabilizes computation like the column scaling of Chapter 1. Residual connections enable gradient flow as in Chapter 9. Position encodings add sequence order through trigonometric functions, while feed-forward layers process attention outputs through standard linear transformations. Each component builds on fundamental principles developed throughout this text, yet their combination transcends the sum of parts.

The computational demands of attention grow quadratically with sequence length, as computing all key-query interactions requires matrix multiplication $QK^T \in \mathbb{R}^{n \times n}$. Modern implementations address this through sparse attention patterns, clever matrix multiplication algorithms, and architectural innovations. Yet the fundamental operation remains matrix multiplication – the same operation that has guided our development from Chapter 1 onward.

The attention mechanism exemplifies how classical mathematics transforms into modern artificial intelligence. Though built from matrices and inner products, it introduces a new level of dynamism where transformations adapt to their inputs. This fusion of timeless mathematical principles with learned adaptation has driven remarkable advances in natural language processing, computer vision, and other domains. Attention unites the precision of linear algebra with the flexibility demanded by intelligence.

13.6 *Representation Learning*

While earlier sections revealed how networks learn to approximate functions, deeper insight emerges from understanding how they discover representations that make such approximation natural. This process – of learning optimal feature spaces rather than merely fitting functions

Foreshadowing: The evolution from hand-crafted to learned representations parallels our journey from analytical matrix factorizations to trained neural networks.

– connects the abstract machinery of previous chapters to the remarkable capabilities of modern artificial intelligence. Like the singular vectors that emerged naturally from matrix structure in Chapter 10, learned representations capture fundamental patterns in data, yet transcend linearity through careful composition of transformations.

Consider first how classical feature engineering differs from learned representations. A hand-crafted feature extractor applies fixed transformations based on domain knowledge: edge detectors for images, frequency analysis for signals, parse trees for text. Though informed by expertise, such features remain limited by human intuition. Neural networks transcend this limitation by learning their own transformations, discovering representational spaces optimally suited to their tasks through end-to-end training.

Definition 13.17 (Representation Space). A *representation space* $\mathcal{H}$ for data $\mathcal{X}$ is a vector space equipped with a learned transformation $f : \mathcal{X} \to \mathcal{H}$ that maps inputs to features. In a deep network with L layers, each hidden layer defines such a space $\mathcal{H}_\ell$ through its learned weights and nonlinearity:

$$\mathcal{H}_\ell = \{\boldsymbol{h}_\ell = \sigma_\ell(W_\ell \boldsymbol{h}_{\ell-1} + \boldsymbol{b}_\ell) : \boldsymbol{h}_{\ell-1} \in \mathcal{H}_{\ell-1}\}$$

where $W_\ell, \boldsymbol{b}_\ell$ are learned parameters and σ_ℓ provides nonlinearity.

This learned structure connects deeply to the *manifold hypothesis* – the principle that high-dimensional data often concentrates near lower-dimensional nonlinear surfaces. While the PCA methods of Chapter 11 discovered optimal *linear* projections of data, deep networks learn *nonlinear* embeddings through composition of transformations. Each layer's weights define a mapping:

$$\boldsymbol{h}_\ell = \sigma(W_\ell \boldsymbol{h}_{\ell-1} + \boldsymbol{b}_\ell)$$

Think: Every image you have ever seen lives in a high-dimensional sub-domain of image space whose complement is of even higher dimension.

that gradually transforms inputs into more structured representations. The nonlinear functions σ allow these mappings to follow curved manifolds, while skip connections and attention mechanisms provide shortcuts through representation space.

Lemma 13.18 (Representation Decomposition). *Let f_ℓ denote the transformation implemented by layer ℓ of a neural network. Each f_ℓ admits decomposition:*

$$f_\ell = \sigma_\ell \circ \tilde{f}_\ell$$

where $\tilde{f}_\ell$ is linear and σ_ℓ provides nonlinearity. The full network learns representations by composing such transformations:

$$f = f_L \circ f_{L-1} \circ \cdots \circ f_1$$

Moreover, each linear component $\tilde{f}_\ell$ induces a decomposition of its domain through the fundamental subspaces:

$$\mathcal{H}_{\ell-1} = \ker(\tilde{f}_\ell) \boxplus (\ker \tilde{f}_\ell)^{\perp}$$

The network learns both the transformations and their associated subspace decompositions through training.

Example 13.19 (Visual Representations). Consider how convolutional networks transform image data through their layers. Early representations capture local patterns like edges and textures – features reminiscent of hand-designed filters. Deeper layers learn increasingly abstract features:

- Layer 1: Edges and orientations
- Layer 2: Textures and simple shapes
- Layer 3: Object parts and complex patterns
- Layer 4: Complete objects and scenes

This hierarchy emerges naturally through training, with each layer learning representations that make the next layer's task simpler. A precise mathematical pattern underlies this emergence: each layer's weights W_ℓ implement linear maps whose singular value decomposition reveals learned feature hierarchies:

$$W_\ell = U_\ell \Sigma_\ell V_\ell^T$$

The right singular vectors V_ℓ capture input patterns while left singular vectors U_ℓ provide the transformed representation basis.

$\diamond$

The power of learned representations becomes clear through the lens of Chapter 12's approximation theory. Just as low-rank matrix factorizations capture dominant patterns in data, deep networks learn features that make complex tasks approximately linear in transformed space. This principle – that good representations linearize otherwise complex problems – appears repeatedly in modern architectures:

Theorem 13.20 (Representation Linearization). *Let $f : \mathcal{X} \to \mathcal{Y}$ be a smooth function between manifolds, and let $\{h_\ell\}_{\ell=1}^{L}$ be the sequence of representations learned by a deep network trained to approximate f. Then under suitable regularity conditions:*

1. *The representation error $\|h_\ell - h_{\ell-1}\|$ decreases with depth*
2. *The nonlinearity of mappings between successive layers decreases*
3. *The final layers implement approximately linear transformations*

Example 13.21 (Word Embeddings). The representation of words as vectors provides a striking example of learned structure. Rather than encoding words through hand-designed features like part of speech or semantic categories, modern language models learn embeddings directly from co-occurrence patterns in text. A word's representation emerges from a learned linear transformation $W \in \mathbb{R}^{d \times v}$ mapping one-hot encodings to dense vectors:

$$e_w = W x_w$$

where $x_w \in \mathbb{R}^v$ denotes the one-hot encoding of word w and $d \ll v$ is the embedding dimension. The rows of W form a learned basis capturing semantic relationships:

1. Similar words cluster together in embedding space
2. Differences between vectors encode analogies
3. Projection onto principal components reveals semantic fields

The embedding matrix W effectively learns a low-dimensional manifold that smoothly captures linguistic structure.

◇

Caveat: The low-dimensional constraint $d \ll v$ is crucial – it forces the network to discover efficient representations rather than memorizing surface patterns.

The study of representation learning connects naturally to the statistical perspective developed in Chapter 11. Just as principal components maximized explained variance in data, learned representations optimize implicit objectives through their architecture and training. Yet unlike the unique optimal projections of PCA, learned representations may vary between training runs or architectural choices. This variability often proves beneficial, as different random initializations can discover complementary feature spaces.

More formally, consider how a deep network transforms its input space through successive representations. Each layer defines a mapping $f_\ell : \mathcal{H}_{\ell-1} \to \mathcal{H}_\ell$ between hidden spaces, culminating in some final representation $\mathcal{H}_L$. The network's power lies not in any individual transformation but in how this sequence of mappings gradually builds structure. Each layer shapes and refines its input until complex patterns emerge from simple foundations.

The mathematics of representation learning reveals a profound principle: intelligence emerges not from raw computation but from learned transformations that render complex patterns simple. This insight – that good representations make hard problems easy – guides our development toward the final synthesis of linear algebra into artificial intelligence. The next section explores how these learned structures ultimately connect back to the fundamental theorem first encountered in Chapter 3, completing our journey from classical mathematics to modern AI.

13.7   Deep Linear Algebra

The mathematics of intelligence reveals itself through fundamental patterns. Though neural networks appear to transcend the structures developed in earlier chapters, their deepest architectures echo and amplify the fundamental theorem first encountered in Chapter 3. At each layer, every transformation between learned representations preserves the essential decomposition that has guided our entire development. This unity – of classical decomposition with learned structure – marks the final transformation of our mathematical journey.

Consider how neural networks shape their representational spaces. Each layer implements not merely a linear transformation followed by nonlinearity, but a learned decomposition reflecting the kernel-image relationship of the Fundamental Theorem. When a layer with weight matrix W maps $\mathbb{R}^n$ to $\mathbb{R}^m$, it induces two complementary splittings:

$$\mathbb{R}^n = \ker(W) \boxplus (\ker W)^\perp \quad \text{and} \quad \mathbb{R}^m = \operatorname{im}(W) \boxplus (\operatorname{im} W)^\perp$$

Unlike the fixed decompositions studied in previous chapters, these spaces emerge through training – the network discovers which directions to preserve and which to collapse. This learned adaptation explains why neural networks often capture patterns that elude hand-designed features: they align their fundamental decompositions with natural structure in data.

Example 13.22 (Learned Decompositions). Consider an autoencoder compressing data through a narrow hidden layer. The encoder $E : \mathbb{R}^n \to \mathbb{R}^k$ and decoder $D : \mathbb{R}^k \to \mathbb{R}^n$ ($k \ll n$) implement dimension reduction through learned transformations whose fundamental spaces acquire profound meaning:

- The *kernel* $\ker(E)$ captures precisely those features deemed irrelevant through training
- The *image* $\operatorname{im}(E)$ provides the learned representation – an optimal k-dimensional summary
- The *cokernel* $\operatorname{coker}(E)$ measures geometric reconstruction loss
- The *coimage* $\operatorname{coim}(E)$ reveals the effective feature space modulo irrelevant variation

The network discovers these spaces by minimizing reconstruction error while respecting the orthogonal structure guaranteed by the Fundamental Theorem. Each space emerges not through mathematical prescription but through learned adaptation to data.   ◇

This principle – that neural networks learn task-appropriate versions of our fundamental decompositions – appears throughout modern architectures. In convolutional networks, each layer's kernel captures local patterns deemed irrelevant while its image preserves task-relevant features. Attention mechanisms implement learned equivalence relations, effectively constructing adaptive quotient spaces by identifying similar tokens. Residual connections create shortcuts through these decompositions, providing direct paths for information flow while maintaining the underlying structure.

The four fundamental spaces that began as abstract concepts in Chapter 3 thus find their deepest expression in how neural networks learn and process information:

- The *kernel* evolves from nullspace to learned invariance, capturing input variations that should be considered equivalent
- The *image* transforms from range to learned manifold, providing representations that meaningfully encode information
- The *cokernel* measures not just algebraic deficiency but learning capacity
- The *coimage* emerges as optimal quotient space, identifying features relevant to the task

Through training, these spaces align naturally with structure inherent in data. The kernel learns task-relevant invariances; the image constrains possible representations; the cokernel guides architecture design; while the coimage captures effective features. Each layer discovers its own version of these fundamental decompositions, transforming our abstract analysis into learned intelligence.

Our exploration of linear algebra concludes as it began: with the profound unity of seemingly distinct concepts. The Fundamental Theorem, which first revealed the complementary subspaces underlying all linear transformations, reaches its fullest expression in the adaptive decompositions of neural networks. In this light, artificial intelligence appears not as revolutionary rupture but as revolutionary refinement – the transformation of timeless mathematical principles into learned form through the structures developed in these pages. Like the singular values that emerged from matrix structure or the eigenspaces that captured asymptotic behavior, the fundamental spaces of linear algebra reveal themselves anew in the learned patterns of modern artificial intelligence.

Convolutional Neural Networks

The synthesis of human and machine vision reveals itself most profoundly in how neural networks process images. When humans look at a photograph, we perceive not mere pixels but meaningful patterns woven from relationships between shapes, textures, and edges. This deep structure — the natural geometry of visual information — guides the design of modern artificial intelligence. Convolutional neural networks achieve their remarkable power by encoding these perceptual principles directly into their architecture, transforming the dense matrix operations of standard networks into specialized transformations that mirror how both human and machine vision parse the visual world.

Consider first how images encode their information. A grayscale image of size $h \times w$ presents itself as a matrix of intensities, yet treating this as an arbitrary array discards crucial spatial relationships. When reshaped into a vector for neural network processing, a modest 256×256 image becomes a point in $\mathbb{R}^{65536}$ — a space of such high dimension that learning arbitrary transformations proves hopelessly inefficient. Yet the patterns we seek to discover — the features that distinguish cats from dogs or tumors from healthy tissue — depend on relationships between nearby pixels rather than arbitrary long-range connections.

This locality principle transforms into mathematical structure through specialized weight matrices that encode spatial relationships. Rather than learning arbitrary transformations, convolutional layers constrain their linear operations to respect the grid-like topology of visual data. A small filter $K \in \mathbb{R}^{k \times k}$ defines a local feature detector that examines each image region in turn — a dramatic reduction in parameters from the $hw \times hw$ matrices of fully connected layers.

The power of this approach emerges from its fusion of biological inspiration with mathematical structure. Just as the visual cortex processes input through localized receptive fields, convolutional networks learn hierarchical feature detectors through carefully constrained linear transformations. Early layers typically discover simple patterns:

- Edge detectors at various orientations
- Color contrast boundaries
- Local texture elements

These combine in later layers to represent progressively more complex features, each emerging from learned compositions of simpler patterns.

Modern architectures enhance this basic structure through innovations that echo ideas from matrix conditioning. Batch normalization layers standardize their inputs across spatial locations:

$$\hat{x}_{ijk} = \frac{x_{ijk} - \mu_k}{\sqrt{\sigma_k^2 + \epsilon}}$$

where k indexes feature channels. Like the scaling techniques that improved matrix conditioning in Chapter 1, this operation stabilizes gradient flow during training. Similarly, residual connections create shortcuts that help networks optimize despite their great depth:

$$Y = X + \sigma(K_2 * \sigma(K_1 * X))$$

This structure, reminiscent of the splitting studied in Chapter 3, provides direct paths for gradient flow while maintaining the spatial structure enforced by convolution.

The remarkable effectiveness of convolutional networks emerges from this fusion of biological inspiration with mathematical principle. Like the low-rank approximations studied in this chapter, they achieve efficiency through careful constraint — not by limiting expressive power, but by encoding known properties of visual data into their matrix operations. Early layers learn localized feature detectors:

- Vertical and horizontal edge detectors
- Center-surround contrast filters
- Simple texture elements

Each represents a specialized linear transformation discovered through optimization yet constrained by architectural design to respect the structure of visual data.

These learned features often exhibit striking universality. Networks trained on natural images reliably discover similar early-layer features regardless of their ultimate task, suggesting these patterns reflect fundamental structure in visual data rather than task-specific artifacts. Later layers specialize more dramatically, but still show recognizable correspondence between networks trained on related tasks.

The success of convolutional networks carries profound implications for representation learning. Their ability to discover effective features automatically, rather than requiring hand-engineered transformations, suggests deeper principles about how structure emerges from data through constrained optimization. This theme — that carefully designed architectures can learn natural representations — appears throughout modern machine learning, from language models to protein structure prediction.

Yet challenges remain. The very architectural constraints that enable efficient learning can sometimes prove limiting:

- Translation invariance may be undesirable
- Long-range dependencies get ignored
- Spatial structure may not match the data

Modern variants like transformers address these limitations through learned attention mechanisms, yet retain the core insight that architectural constraint enables efficient learning.

The mathematical principles underlying convolutional networks extend far beyond computer vision. Their fundamental insight — that local structure matters and should be architecturally encoded — has inspired specialized networks for many domains:

- Graph convolution for molecule modeling
- Spherical convolution for climate data
- Temporal convolution for time series

Each adapts the basic mathematics to respect domain-specific structure while maintaining the core principle that architectural constraint enables efficient learning.

Beyond their practical impact, convolutional networks exemplify how classical mathematics transforms into modern machine learning through careful engineering. The linear transformations studied throughout this text provide the foundation, while thoughtful architectural constraints enable efficient learning of natural representations. As artificial intelligence continues its rapid advance, this principle — that structure enables learning — will likely prove increasingly valuable.

Large Language Models & Algebraic Reasoning

The ultimate synthesis of linear algebra into intelligence reveals itself most profoundly in how vast neural networks process language. When humans read text, we perceive not mere sequences of words but intricate webs of meaning woven from grammar, context, and knowledge. This deep structure — the natural geometry of linguistic information — guides the design of modern language models. Through careful composition of attention mechanisms and learned representations, these systems transform the mathematics developed throughout this text into apparent understanding, achieving capabilities that would have seemed miraculous mere years ago.

Consider first how language encodes its meaning. Each word or subword token maps to a learned vector in high-dimensional space — a point in $\mathbb{R}^d$ where typically $d \approx 1024$ or larger. Yet treating these embeddings as arbitrary vectors discards the profound structure of language. When processing the phrase "the cat sat on the mat," a language model must somehow capture not just individual word meanings but their grammatical relationships, semantic roles, and broader context. The patterns we seek to model — the features that distinguish questions from statements or detect subtle implications — emerge from complex interactions between these representations.

This relational structure transforms into mathematical form through the attention mechanisms studied earlier in this chapter. Rather than applying fixed transformations, each layer computes dynamic relationships between all pairs of tokens:

$$A_{ij} = \frac{q_i^T k_j}{\sqrt{d}} \quad \text{and} \quad \tilde{A}_{ij} = \frac{\exp(A_{ij})}{\sum_k \exp(A_{ik})}$$

The resulting attention weights $\tilde{A}$ implement learned equivalence relations, effectively constructing quotient spaces that group tokens by their contextual roles. Like the fundamental decompositions first encountered in Chapter 3, these dynamic quotients organize information flow through the network.

The power of this approach emerges from its fusion of mathematical elegance with practical effectiveness. Just as the quotient spaces of earlier chapters revealed structure by identifying equivalent elements, attention mechanisms discover contextual patterns by learning which tokens should interact. Early layers typically capture surface properties:

- Basic syntactic relationships
- Local word associations
- Simple referential patterns

These combine in later layers to represent increasingly abstract relationships, each emerging from learned compositions of simpler patterns.

The sheer scale of modern language models — with hundreds of billions of parameters — reveals new principles beyond those visible in smaller systems. Like massive crystals whose microscopic structure gives rise to emergent properties, these networks develop capabilities that seem to transcend their basic components:

- Few-shot learning from textual examples
- Zero-shot transfer to novel tasks
- Apparent reasoning and inference

Yet their foundation remains the matrix operations studied throughout this text — attention computing relevance scores through inner products, feed-forward layers implementing learned transformations, layer normalization stabilizing gradients.

This emergence of intelligence from scale illuminates profound questions about representation and learning. The vectors and matrices processing each token encode not just word meanings but fragments of knowledge, reasoning patterns, and implicit skills. Like the singular vectors that emerged naturally from matrix structure in Chapter 10, these learned representations capture fundamental patterns in language — but now discovered through optimization rather than analysis.

Consider how a language model processes the prompt "Complete the pattern: 2, 4, 6, ...". Each token's embedding transforms through layers of attention and feed-forward computation:

$$h_\ell = \text{FFN}_\ell(\text{Attention}_\ell(h_{\ell-1})) + h_{\ell-1}$$

The residual connections, reminiscent of the splitting studied in Chapter 3, allow information to flow directly while attention mechanisms extract relevant patterns. Somehow from these mathematical operations emerges the ability to recognize and continue sequences — an apparently simple capability that requires sophisticated processing of both language and number concepts.

Yet challenges remain. The very architectural features that enable this emergence also impose limitations:

- Attention scales quadratically with sequence length
- Models can produce plausible but incorrect responses
- The basis for their capabilities remains somewhat opaque

These limitations remind us that while current systems achieve remarkable results, they represent steps along a path rather than its endpoint.

The mathematical principles underlying language models extend far beyond text processing. Their fundamental insights — that attention enables dynamic information flow, that scale enables emergence, that careful architecture design channels learning — have inspired applications across domains:

- Protein structure prediction
- Scientific discovery
- Code generation and analysis

Each adapts the basic mathematics to new contexts while maintaining the core principle that structured computation enables complex behavior.

Historical Note: The term "attention" emerged from neural machine translation, where these mechanisms first showed their power. Their true potential became apparent only with the transformer architecture in 2017, leading directly to modern language models.

Beyond their practical impact, large language models exemplify how classical mathematics transforms into modern artificial intelligence through careful engineering. The linear transformations studied throughout this text provide the foundation, while thoughtful architectural choices enable efficient learning of natural representations. As these systems continue their rapid advance, this principle — that structure enables intelligence — will likely prove increasingly valuable.

The journey from basic linear algebra to language models tracks our text's broader theme: how timeless mathematical principles transform into practical tools through careful development. These systems forge understanding from the raw material of computation. Their success suggests not that traditional mathematics has been superseded, but that it finds new expression through the structures we create. In this light, large language models appear not as breaks from classical principles but as their natural evolution — the transformation of linear algebraic operations into apparent intelligence through principled composition at scale.

Exercises: Chapter 13

1. Let $W \in \mathbb{R}^{m \times n}$ and $b \in \mathbb{R}^m$ define a neural network layer computing $h = \max\{0, Wx + b\}$. Show that without the ReLU activation $\max\{0, \cdot\}$, this reduces to transformations from Chapter 3. Then for $n = m = 2$, explicitly describe the regions of input space where this transformation is linear, illustrating with a diagram.

2. For the softmax function $\sigma_i(x) = e^{x_i} / \sum_{j=1}^n e^{x_j}$: Prove that outputs sum to 1 for any input x; derive the derivative matrix entries $\partial \sigma_i / \partial x_j = \sigma_i(\delta_{ij} - \sigma_j)$; and explain why adding a constant to all inputs leaves outputs unchanged.

3. Consider layer normalization $\hat{h} = \gamma(h - \mu)/\sqrt{\sigma^2 + \epsilon} + \beta$ where μ, σ^2 are activation mean/variance and γ, β are learned parameters. Show the output has mean β and variance γ^2, relating this to the column scaling of Chapter 1. How does this help control weight matrix condition numbers?

4. For a word embedding matrix $W \in \mathbb{R}^{d \times v}$ mapping vocabulary to d-dimensional vectors: Prove its row space dimension is at most d; connect this to Chapter 12's low-rank approximations; and explain what similar contexts imply about corresponding rows of W.

5. The residual connection $h_{\text{out}} = h_{\text{in}} + F(h_{\text{in}})$ adds input directly to transformed output. Express $\partial h_{\text{out}} / \partial h_{\text{in}}$ in terms of F's Jacobian and explain how this helps deep network training, connecting to the conditioning concepts of Chapter 4.

6. Let $E : \mathbb{R}^n \to \mathbb{R}^k$ and $D : \mathbb{R}^k \to \mathbb{R}^n$ be an autoencoder's encoding/decoding maps with $k < n$. Identify E's four fundamental subspaces, explaining how $\ker(E)$ relates to discarded information and $\operatorname{coker}(E)$ to reconstruction error. Connect to the theory developed in Chapter 3.

7. For attention weights $A_{ij} = \exp(q_i^T k_j / \sqrt{d}) / \sum_l \exp(q_i^T k_l / \sqrt{d})$: Prove each row sums to 1; analyze behavior as $d \to 0$ and $d \to \infty$; and relate to the stochastic matrices from Chapter 9.

8. Multi-head attention combines h parallel computations through $\text{MultiHead}(Q, K, V) =$

$W_O[\text{head}_1; \ldots; \text{head}_h]$. Show this reduces to standard attention when $h = 1$ and explain why multiple heads might better capture different relationship types. Connect to how SVD's singular vectors capture different patterns.

9. Consider a neural network applying L layers of $\boldsymbol{h}_{\ell+1} = \sigma(W_\ell \boldsymbol{h}_\ell + \boldsymbol{b}_\ell)$. Show that without activations this reduces to matrix multiplication; explain how nonlinearity enables curved manifold approximation; and connect to representation learning from Section 13.6.

10. (Challenge) For a transformer using positional encodings $\{\boldsymbol{p}_1, \ldots, \boldsymbol{p}_n\}$ with sequence $\{\boldsymbol{x}_1, \ldots, \boldsymbol{x}_n\}$: Prove attention can detect relative positions through inner products; demonstrate how sinusoidal encodings facilitate this; and connect to Chapter 5's orthogonality concepts.

11. (Challenge) Let f be a neural network with more parameters than training points $\{(\boldsymbol{x}_i, y_i)\}_{i=1}^{n}$, trained to minimize mean squared error. Show it can achieve zero training error by analyzing the associated Jacobian's rank. Relate this to the universal approximation theorem from Section 13.1.

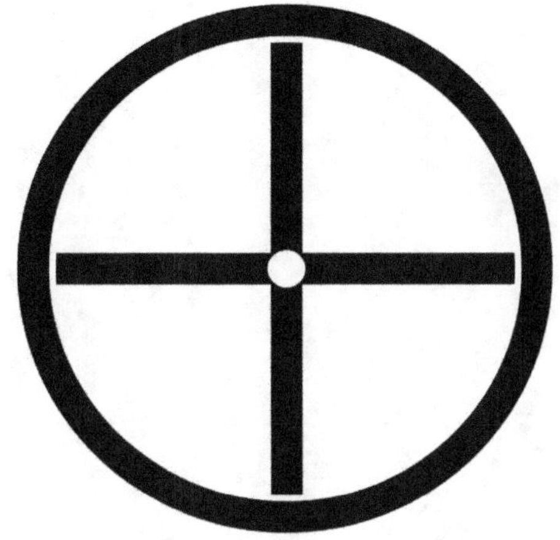

& these again surrounded by
Four Wonders of the Almighty Incomprehensible
Pervading all amidst & round about
Fourfold each in the other reflected
They are named Life's in Eternity
Four Starry Universes going forward
From Eternity to Eternity

ABOUT THE AUTHOR

Robert Ghrist (Ph.D., Cornell, Applied Mathematics, 1995)
is the Andrea Mitchell PIK Professor of Mathematics and
Electrical & Systems Engineering at the University of Pennsylvania.
He is a recognized leader in the field of Applied Algebraic Topology,
working in networks, robotics, signal processing, data analysis,
optimization, and more. He is an award-winning researcher,
teacher, and expositor of Mathematics and its applications,
currently serving as the Associate Dean of Undergraduate Education
in the School of Engineering & Applied Sciences at Penn.

He is the author of several books, such as:
Elementary Applied Topology and the *Calculus Blue Guide*;
as well as the creator of YouTube video series, including
Calculus BLUE, Calculus GREEN, & Applied Dynamical Systems.

In his spare time he publishes mathematical art
and animation under the moniker *colimit*.